BRICKWORKS

AUSTRALIA

australbricks®

australmasonry®

bristileroofing®

southern cross cement®

BOWRAL BRICKS

DANIEL ROBERTSON

NUBRIK

GB Masonry

URBANSTONE®

Terraçade

CAPITAL BATTENS

NORTH AMERICA

Glen-Gery

BRICKWORKS® SUPPLY CENTER

BRICKWORKS® DESIGN STUDIO

INTERNATIONAL BRANDS

La Escandella ROOFING THE WORLD

S.ANSELMO vivere la differenza

la paloma cerámicas

CERAMICHE KEOPE

The job of the architect today is to create beautiful buildings. That's all.

Philip Johnson

Materiality

Brick in contemporary university architecture

North America edition

Editor & curator
Ron Ringer

Materiality: brick in contemporary university architecture (North America edition) is the fourth in a series of books that explores the use of brick in contemporary architecture.

As with previous volumes, this latest publication offers case studies and illustrated technical essays that provide a high level view of how the architect and designer imagined and executed their project. Richly illustrated with photographs and drawings, the intention is to draw the reader into the world of outstanding design and the artful use of brick.

The decision to create a publication showcasing campus and community architecture projects is a natural progression for Brickworks, the parent company of Glen-Gery, one of its US subsidiaries. Many of the projects feature bricks made by Glen-Gery, as well as those manufactured by other producers which are distributed through the Brickworks distribution network. What is important to understand is the universal appeal of brick, regardless of its point of origin.

This edition of *Materiality* also includes feature articles that explore the technical nature of how bricks are being incorporated in buildings in new and artful ways. Technical features on brick bonds, arches, shapes, and the largest and most modern brick plant in the world are intended to provoke discussion and interest among those in the industry, the general reader, and students of architecture and design. These so-called technical features continue to be an integral part of the ongoing debate about materiality, form and function in the built environment.

Ron Ringer is the published author of numerous commissioned works, including business histories and educational textbooks. He is also a business communications consultant, a teacher of academic writing, and an independent historian who has worked extensively across public and private sector organisations and educational institutions. Ron established Syntax Writing Services Pty Ltd in 1991 following a successful career as a teacher in the UK (Liverpool) and Australia (Sydney). Since leaving the teaching profession in 1989 he has worked with many of Australia's leading banks, tertiary institutions, and corporations to provide communication and consulting services on simplifying business communication and documentation.

Ron Ringer's previous works of history include: *JANDS: An Australian Story* (2023); *Industrial Arts: A history of practical education in NSW schools, 1820-2020* (2020); *Materiality 2021: brick and block in contemporary architecture* (2021); *The Material City: density and design in contemporary Australian architecture* (2019); *A history of Twilight Aged Care, 1915-2015* (2017); *Summoned by bells: A history of St Jude's, Randwick, 1865-2015* (2016); *Materiality: brick and block in contemporary Australian design* (2015); *Along the Wayside: 50 years of The Wayside Chapel* (2014); *The Spirit of SMEC: Snowy Mountains Engineering Corporation* (2012); *The Brickmasters: 1788-2008* (2008); *A History of Elgas Limited* (2004). Ron has also written numerous textbooks on the subject of HSC Modern History (1987-2018) for Pascal Press, Macmillan Educational Australia and Pergamon Press.

First published in Australia in 2024
by Brickworks Building Products North America Pty Ltd
ABN: 93 000 002 979

For further information, contact Brickworks Limited, Wallgrove Road, Horsley Park, NSW 2164, Australia. T: +61 2 9830 7700 or Ron Ringer (editor and curator of the Materiality series of publications, T: +61 412 560 151).

National Library of Australia Cataloguing-in-Publication entry

NATIONAL LIBRARY OF AUSTRALIA

A catalogue record for this work is available from the National Library of Australia

Title: Materiality: brick in contemporary university architecture (North America edition) / Ron Ringer

ISBN 978-0-9941654-1-1 (hardback)

Other Creators/Contributors:

Editor, curator, author: Ringer, R. E. (Ronald E.), 1953.
T: +61 412 560 151 , E: ronringer2013@gmail.com
www.syntaxwriting.com.au

Design: Taninka Visuals, E: tanya@taninka.com.au

Printing: The Standard Group
433 Pearl Street, Reading, PA 19602
www.standardgroup.com

Front cover Zeppos Residential College, David M. Schwarz Architects. Image: Steve Hall, Hall+Merrick+McCaugherty.

Page 003 NYS Equal Rights Heritage Center, nARCHITECTS. Image: James Ewing/OTTO.

Page 005 Institute for Data Science, Kennedy & Violich Architecture. Image: John Horner, courtesy of KVA.

Page 006 Pauli Murray College and Benjamin Franklin College Robert A.M. Stern Architects. Image: Peter Aaron/OTTO.

Page 008 Sid Richardson Residential College, Barkow Leibinger. Image: Iwan Baan.

Page 011 The Owsley Brown II History Center, de Leon & Primmer Architecture Workshop. Image: de Leon & Primmer Architecture Workshop.

Introduction
Lindsay Partridge

I am delighted to introduce the latest book in the *Materiality* series which explores how architects and designers are using brick in campus and community architectural projects across the United States of America. All of the firms that contributed case studies are leaders in their field, outstanding in their craft and exemplars of innovation in the use of this beautiful, enduring building material—brick.

The work of designers is to enrich our enjoyment, enhance our understanding and build our experience of the built environment. The projects in this book represent only a fraction of the tens of thousands produced by professionals over the last two decades in particular.

The illustrated technical essays and features are equally important in that they demonstrate the enormous versatility of brick, and seek to remind us of skills that continue to be in great demand. Modern production methods are revealed in detail in a format that we consider to be a first.

The brick manufacturing industry and the vitality of our distribution networks with their local and statewide connections are fundamental to our existence. The people employed across this far-reaching endeavor form an integral part in the process of production, sales and marketing, delivery and construction.

All of us play an important part in a business that grows, year on year. Brickworks and its US subsidiaries will continue to be at the forefront.

In these pages you will find outstanding examples of architecture where brick is vital, visible and engages our imagination. I commend this book to you, and also those who have worked to bring us yet another remarkable publication.

Enjoy the read.

Lindsay Partridge, AM
Managing Director, Brickworks

Recognition
Ron Ringer

As with all previous books in the *Materiality* series the challenge has been to achieve balance—in the selection of outstanding designers, outstanding projects, and unusual ideas that arise from the use of brick. Each and everything contribute to our understanding of the places in which we work, study and live.

Tertiary—higher—education helps to shape the minds of people of all ages, which makes 'place' all the more important. Inspiring architecture that creates fit-for-purpose learning and teaching facilities is the subject of this volume. My intention has been to inform and educate the widest possible audience, from the student who enters a building that conveys an immediate sense of awe, to the designer who finds remarkable the work of a colleague in a far off town or city campus. There is ample room, too, for the prospective client who discovers inspiration in a particular building to be found within these pages.

Thank you, Brett Ward, for being such a great colleague in the making of these publications, and thank you, Lindsay Partridge, for continuing to support our work. A special thanks to Stephen Varady, always helpful and ever the gentleman, for his insights during the selection process. Thanks in good measure to our colleagues at Glen-Gery in the USA. They know who they are, especially the team at Brickworks Design Studio on New York's 5th Avenue who were so welcoming.

Without my design colleague and collaborator, Tanya Deenick (and her associate, Anna Carden) these books simply would not work—or at least work as well. From me, there's huge respect for their brilliant work. Richard Blair's eagle eye at various stages of proofreading helped us cross the finish line.

Finally, to my wife, Jaana, who is always there.

Ron Ringer, editor & curator

Foreword
Jeremy Edmiston

Working as an architect in the twenty first century on Planet Earth, home to nearly 8 billion inhabitants, often entails collaboration with neighbors. Architects featured in this edition of Materiality have discovered the enduring value of brick in this collaborative endeavor. Neighborhoods, streets, and campuses exhibit diverse characteristics, with the preservation of history becoming a cultural ambition enforced through heritage boards, landmark reviews, and preservation laws. As evident in the projects discussed here, materiality serves as a crucial tool for reconciling contemporary building functions and technologies with the diverse age and purpose of neighboring structures.

Brick, an ancient building material dating back to 7,700 BCE, is crafted from clay—an abundant and naturally occurring substance. Traditionally, it has been sourced locally, grounding structures in their specific locales. In New York City, for instance, many bricks were sourced from the nearby Hudson River Valley, contributing to the city's exponential growth from 500,000 people in 1850 to over five and a half million by 1920.

While brick is a diminutive building block, it has been employed in myriad ways. In the twentieth century, structural steel gained prominence in North America, leading architects to adapt brick to protect steel frames from weather and fire. Contemporary examples showcase innovations like brick rain-screens, which shed water from building surfaces, relying on backup constructions for environmental and structural integrity.

Current brick construction often employs a single wythe (a single unit of depth) for the outer layer or internal finish, with visible brick patterns often based on historic bonding techniques that were used for tying multiple wythes together. Patterning emerges as a key characteristic, animating surfaces and creating a dynamic play of light when exposed to sunlight.

In recent projects, pattern extends beyond flat surfaces, incorporating three-dimensional elements through brick corbelling. Originally used for detailing specific architectural features, corbelling is now deployed across entire surfaces, adding texture and depth. Some designs use specially shaped bricks or polychromy—multiple colored bricks—further expanding the possibilities of this ancient technique. Used now with a freedom that would no doubt excite the nineteenth century English architect William Butterfield, polychrome brickwork has a history all of its own.

These timeless techniques, bridging the past and present, facilitate a dialogue not only with architectural elements but also with the hands that laid the bricks. While digital advancements have played a role in the resurgence of brick surfaces, the craft and precision required in its application align with both digital and hand-drawn design practices.

The focus of this book is on constructions commissioned by institutional clients with a commitment to the longevity of their campuses. Brick, known for its durability and adaptability, signals longevity and serves as a low-maintenance, long-lasting building system. Despite the energy costs associated with firing clay, the permanence of brick encourages use and re-use, aligning with sustainable building practices.

In New York City, many nineteenth and early twentieth century buildings faced with brick still stand, their endurance evident in the proliferation of scaffolding for maintenance. These structures, adapted and repurposed over time, play a vital role in the city's civic life, increasingly recognized and preserved as landmarks that contribute to its rich history.

Jeremy Edmiston, Principal
System Architects LLC

Contents

Materiality
Case studies

DePaul University Art Museum

Antunovich Associates

Location Chicago, Illinois
Function Art Museum
Client DePaul University
Completion date 2011
Architect Antunovich Associates
General Contractor Bulley & Andrews Construction
Structural Engineer Tylk Gustafson Reckers Wilson Andrews, LLC (TGRWA)
Civil Engineer Spaceco Inc.
MEP/FP Salas O'Brien (formerly WMA Consulting Engineers)
Photography Antunovich Associates

Following the renovation of the Fullerton CTA Station, the DePaul University Art Museum marked the initial phase of a series of campus enhancements for the University's Lincoln Park Campus. Positioned between the bustling CTA elevated tracks and the Fullerton station to the west, and the neighboring historic McCormick Rowhouses to the east, the Art Museum was conceived to serve as a buffer, shielding these residences from the noise and commotion of the CTA station. Simultaneously, it harnessed the convergence of pedestrian activity and public transit accessibility.

Within its compact footprint, the Art Museum accommodates galleries, presentation spaces, and collections study areas across three levels, offering versatile exhibition spaces and opportunities for engagement with the museum's collections. Beyond fostering visual cultural exchanges between the University and the Lincoln Park community, the Art Museum also facilitates the educational goals of DePaul's Museum Studies program. The ground and second floors house five gallery spaces, while the third floor encompasses collections study and exhibition preparation

ART MUSEUM
935
DePAUL UNIVERSITY

Traditional and innovative masonry detailing further enlivens the brick and stone façade.

2

areas, in addition to administrative offices. Transitioning between these levels is achieved through an ornamental circulation stair characterized by terrazzo plank treads and a glass and metal railing system. An oversized service elevator caters to efficient transportation of artwork to upper-level galleries as well as object preparation and study rooms. A protective canopy over the side entry doors offers cover during collection loading, shielding against the elements.

Architectural aesthetic The design of the Art Museum's exterior pays homage to the architectural style of the nearby McCormick Rowhouses, seamlessly extending the brick and limestone aesthetic tradition employed by DePaul University in response to Lincoln Park's distinctive ambience. By melding red brick walls with buff limestone accents and incorporating a series of blind recesses, the building echoes the scale of the surrounding neighborhood. A blend of conventional and innovative masonry detailing further animates the brick and stone façade. The ground floor openings that face Fullerton Avenue feature self-supporting arches – a three-centered arch crowns the central window, flanked by flat arches on either side. The corbelled parapet, reminiscent of a prominent local architectural element, graces the building, achieved through careful manipulation of the veneer brick within the cavity wall system.

3

1 Exterior façade (north) of Art Museum, from West Fullerton Avenue.

2 Exterior façade showing detail of monumental arched window and entrance, looking east along West Fullerton Avenue.

3 Ground floor reception area looking north.

4 Interior art installation in second floor Gallery 3, with bay window at right.

5 Exterior view of the east façade of the Art Museum, with canopy.

Exterior cladding encloses a highly engineered internal envelope structure to address fundamental museum programming concerns.

4

5

Practicality and function The veneer brick exterior clads an engineered internal structure, designed to address the foundational concerns of museum programming. Caisson footings, reinforced and grouted CMU block exterior walls, and precast plank floors work in unison to isolate the gallery spaces from the noise and vibrations emanating from the nearby CTA train station and Fullerton Avenue. Fenestration is strategically positioned on the exterior to limit the penetration of light (both visual and UV), sound, and vibrations into the gallery spaces. A defining bay window, the sole glazed opening on the west wall, provides views along the CTA tracks and towards the Chicago skyline. Windows are also strategically incorporated to provide natural illumination within the ground floor reception area, the second-floor presentation room, the third-floor offices, and the vertical circulation vestibule on the east façade. The punched window openings feature triple-pane glazing with Low-E coatings, while the monumental arched window facing Fullerton Avenue is embellished with tempered laminated glass.

North elevation

East elevation

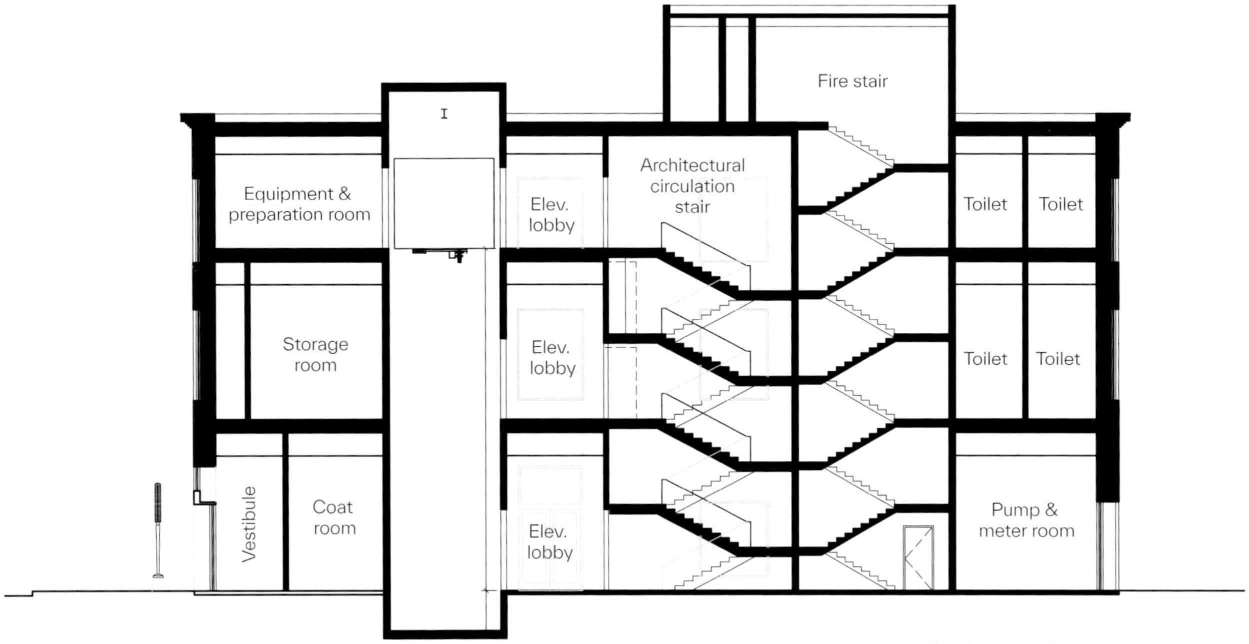

Stairs section - looking east

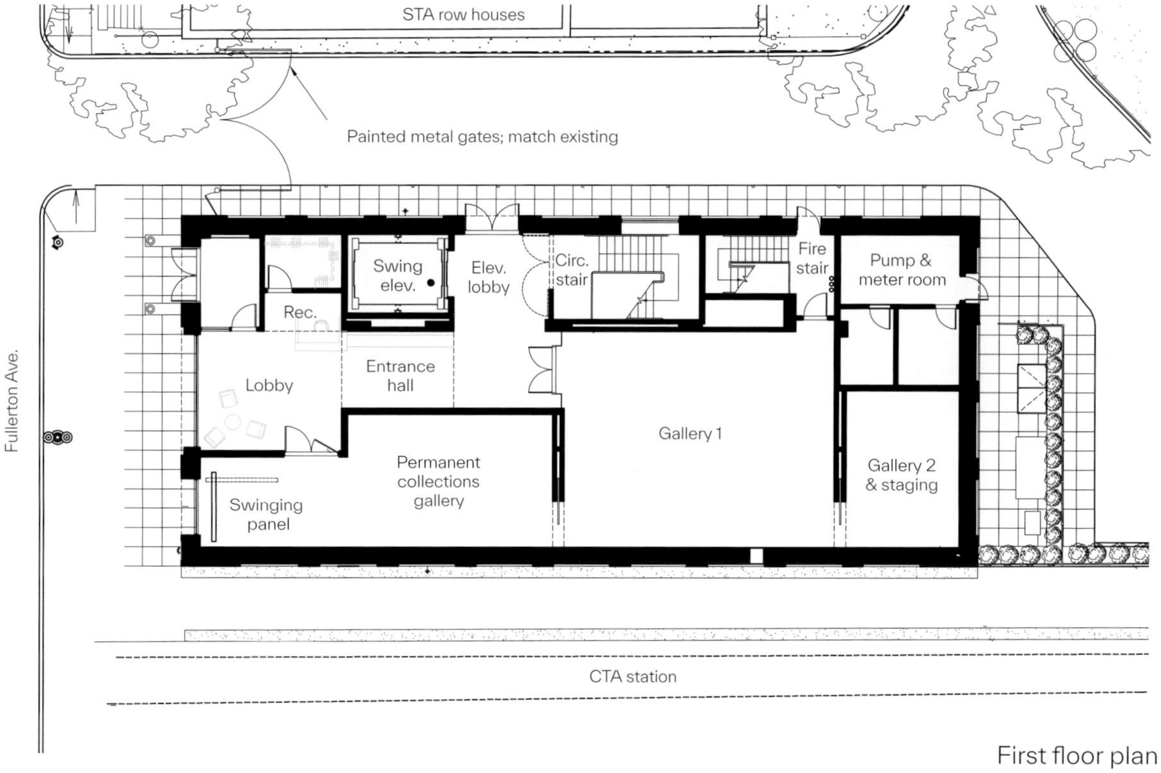

First floor plan

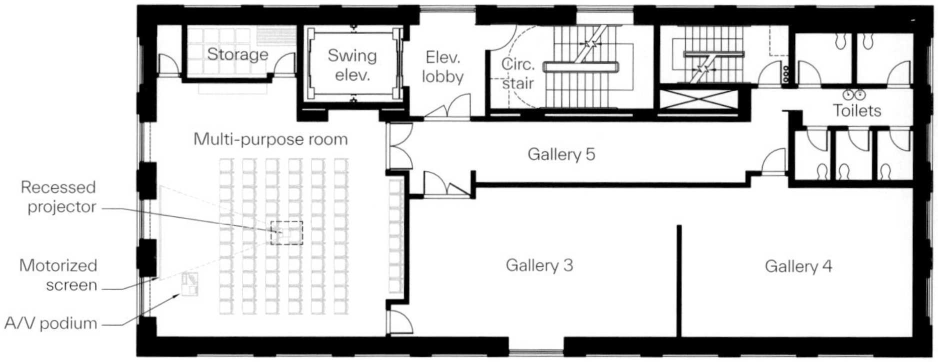

Second floor plan

Holtschneider Performance Center

Antunovich Associates

Location Chicago, Illinois
Function Performance Center
Client DePaul University
Completion date 2018
Architect Antunovich Associates
Sustainability LEED Gold Certified
General Contractor Bulley & Andrews Construction
Structural Engineer CS Associates
Acoustical Consultant Kirkegaard
Theatre Planners & Lighting Designers Schuler Shook
MEP/FP Salas O'Brien
Photography balloggphoto.com

The Holtschneider Performance Center at DePaul University School of Music is a meld of sound and craftsmanship. Carefully designed and expertly executed, this architectural project embraces cutting-edge acoustics that offer an ideal arena for music scholars to both learn and showcase their talents. Beyond its functional excellence, the center sets the stage for a purposive educational experience that can be enjoyed in the spacious gathering zones and fenestration designed to bathe the spaces in natural light.

Set within this new musical epicenter are the Mary Patricia Gannon Concert Hall, accommodating 505 attendees, the intimate Murray and Michele Allen Recital Hall with seating for 140, the Brennan Family Recital Hall that accommodates 80, and the Mary A. Dempsey and Philip H. Corboy Jazz Hall with seating for 75. The structure also houses state-of-the-art rehearsal spaces, a suite equipped with the latest in sound recording technology, a dedicated box office, a welcoming atrium café, and versatile classrooms that can easily adapt to different teaching methodologies. Furthermore, individual practice rooms offer sanctuaries for students. Every performance space is designed with customizable acoustic features, including battered walls, strategically isolated structural components, and specialized

1
DEPAUL UNIVERSITY
SCHOOL OF MUSIC
EMERGENCY
2018

2

surface treatments. The project's emphasis on sustainable construction is evident through the incorporation of local, recycled, salvaged, low-emission, and renewable materials.

The performance center hosts an array of venues, from grand concert halls to recital spaces and jazz stages. Thoughtfully positioned student rehearsal rooms, practice areas, and classrooms occupy the upper levels, optimized for functional and acoustic congruence with the towering performance areas below. The Schaefer College Hall Lobby, a three-story atrium, greets visitors at the entrance, doubling as a connecting thoroughfare that deftly organizes the building's spatial layout. This 'interior street' serves as an inviting pathway, bridging the center with DePaul's Lincoln Park Campus, fostering inclusivity and accessibility.

3

4

The project incorporates local, recycled, salvaged, low-emitting and renewable materials wherever possible.

5

An organic bond The Holtschneider Performance Center creates a harmonious connection with its surroundings, becoming an integral part of the community it serves. Its exterior, primarily crafted with brick and adorned with stone panels and horizontal coursing, references the historic McCormick Rowhouses nearby. Glass and stone, thoughtfully interspersed throughout the façade, modulate the building's visual weight and lend an animated aura along Halsted Street. The architectural scale gracefully navigates between the towering concert hall's demands and the surrounding residential context through strategically placed setbacks.

Beyond its architectural congruence, the center holds a profound cultural sway over the community. Since its inauguration, DePaul's Community Music Program has witnessed a marked upswing in enrollment, now embracing over 700 students from the Lincoln Park neighborhood and beyond. This program encompasses various student cohorts, including children's choirs and community choruses, all of whom are poised to leverage the center's four performance spaces along with the many practice and rehearsal quarters offered by this new, expansive facility.

1 West exterior, entrance detail.
2 Murray And Michele Allen Recital Hall.
3 Schaefer College Hall Lobby.
4 Mary Patricia Gannon Concert Hall
5 West exterior, facing Halsted Street.
6 Brennan Family Recital Hall.

6

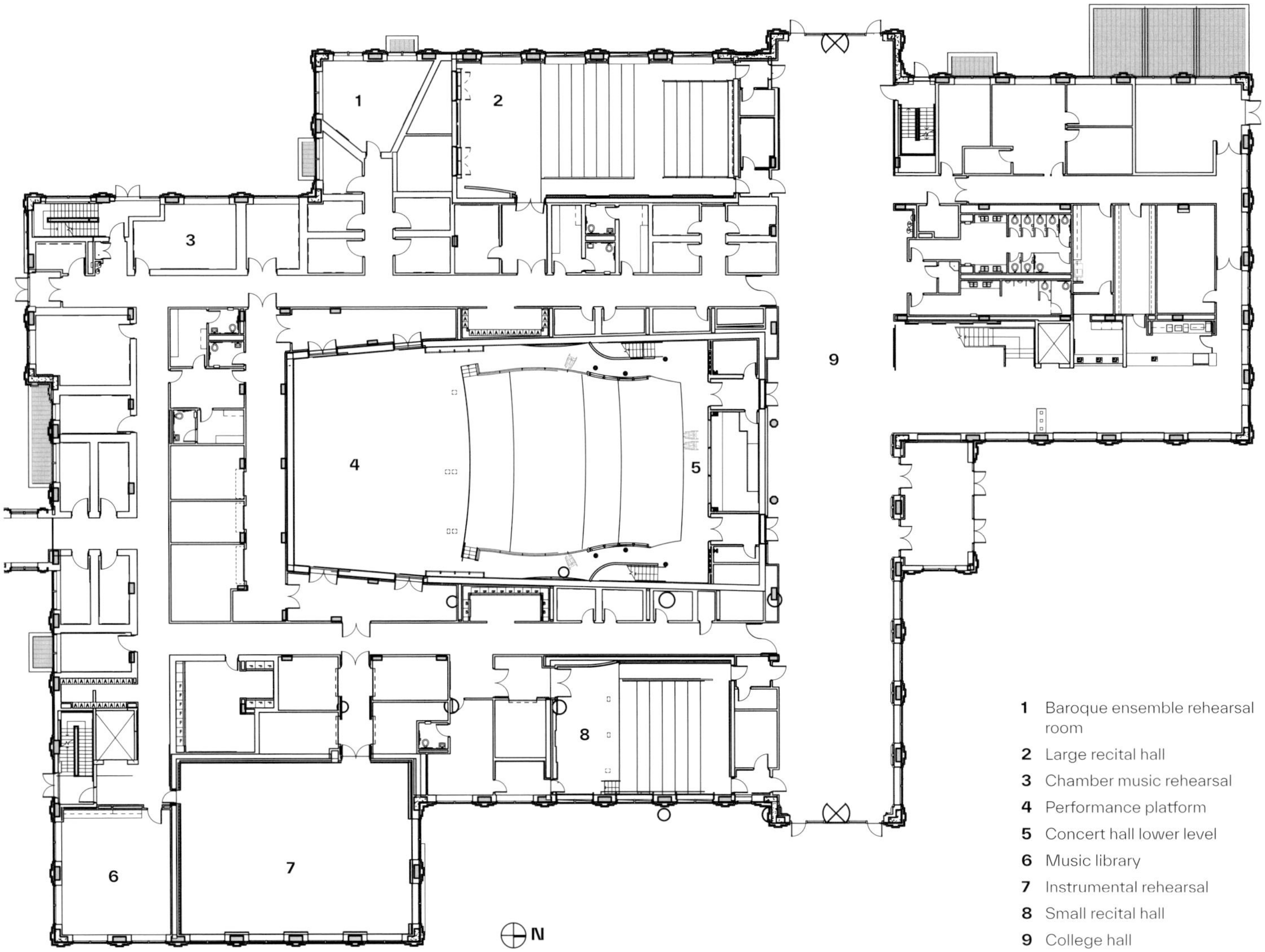

Level one

1 Baroque ensemble rehearsal room
2 Large recital hall
3 Chamber music rehearsal
4 Performance platform
5 Concert hall lower level
6 Music library
7 Instrumental rehearsal
8 Small recital hall
9 College hall

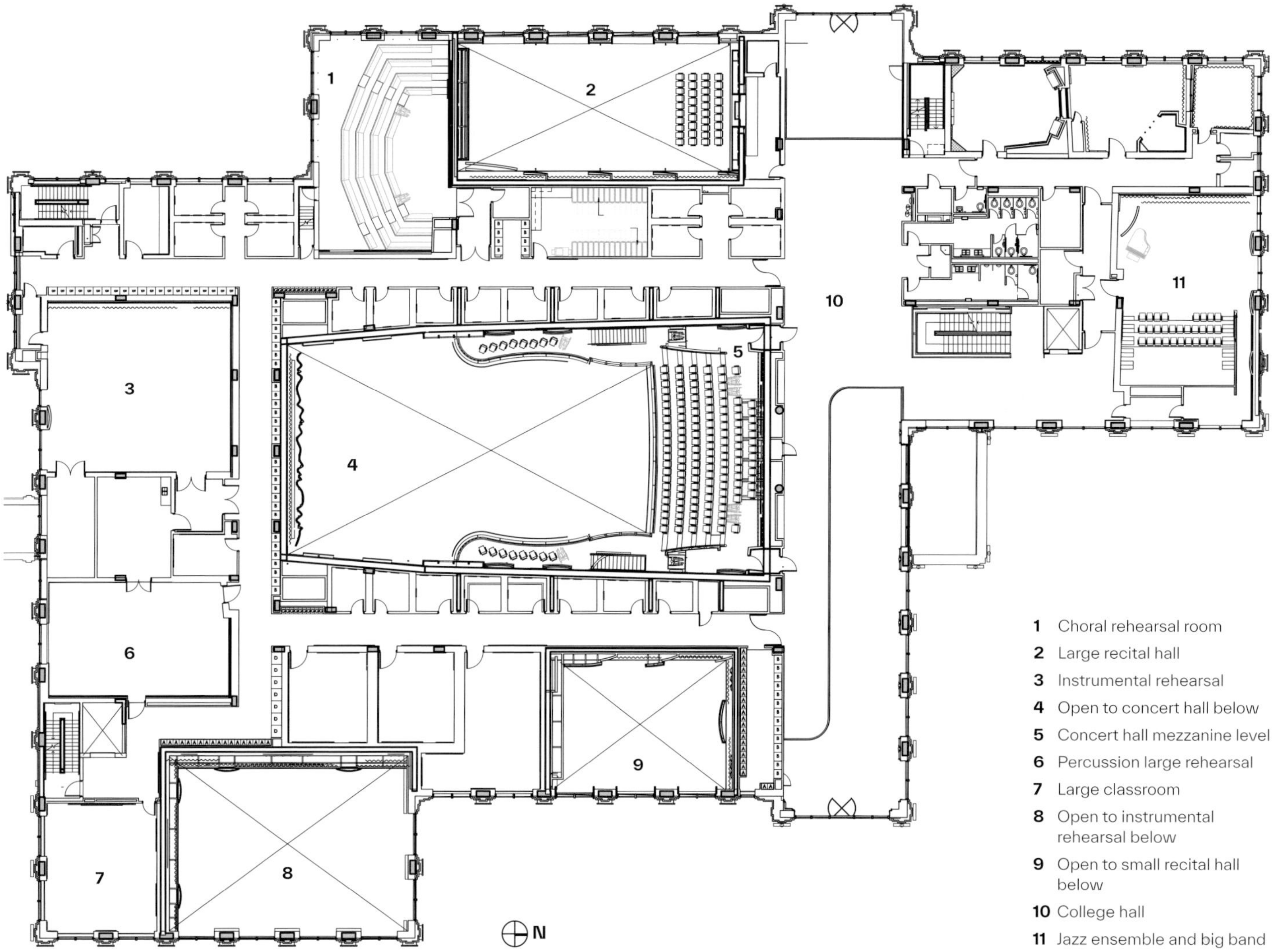

1 Choral rehearsal room
2 Large recital hall
3 Instrumental rehearsal
4 Open to concert hall below
5 Concert hall mezzanine level
6 Percussion large rehearsal
7 Large classroom
8 Open to instrumental rehearsal below
9 Open to small recital hall below
10 College hall
11 Jazz ensemble and big band

Level two

Sid Richardson Residential College
Barkow Leibinger

Location Houston, Texas
Function Residential College
Client Rice University
Completion date 2021
Architect Barkow Leibinger, Berlin
Landscape The Office of James Burnett, Houston
Structural Engineer Knippers Helbig, New York + Ensight Haynes Whaley, Stafford
Technology & Security 4b Technology Group, Houston
MEP I.A. Naman+Associates, Houston
Acoustical Engineer Jaffe Holden, Houston
Civil Engineer Walter P. Moore, Houston
Architect of Record/LEED consultant Kirksey Architecture, Houston
Geotechnical Engineer Ulrich Engineers
General contractor Anslow Bryant Construction LTD, Houston
Photographers Iwan Baan, Amsterdam
Joe Aker/Aker Imaging, Houston

1

The new Sid Richardson Residential college marks an important milestone in Rice University's expansion in Houston. Rice University has a history of supporting various architectural styles, from its classical origins in the 1910 masterplan by Cram, Goodhue, and Ferguson, to modernism in the 1950s and 1960s, post-modernism in the 1980s, and more recent innovative structures by architects like Michael Malzahn, Thomas Phifer, and artist James Turrell. This new college caters to a diverse social and cultural community, housing student dormitories, learning spaces, recreational facilities, dining areas, offices, and accommodations for the magister, consisting of three structures of different heights.

The arrangement of a 12-story and a 5-story residential building, along with a 2-story events building featuring a terrace, creates a connection between the urban and campus environments. The ensemble bridges the gap between the tall medical centers on Main Street and the tranquil Sid Rich Quad. The project offers a total of 312 beds, counselor apartments, and an apartment for a teaching assistant. The communal spaces are situated on the ground floor, seamlessly connecting all three components. The centerpiece is a column-free dining hall with seating for 300. Its distinct feature is a fan-like roof design that encourages social interactions.

The lower-rise section accommodates workshops and event spaces. A rooftop terrace acts as a bridge between living quarters and communal areas. A 'veranda' concept, developed in collaboration with students and staff, serves as a social hub. Photovoltaic modules integrated into triangular

2

An incubator for architecture, Rice has a long history of supporting forward-looking architecture from its classical origins.

3

The brick is a mixture of warm blends and textures: a zigzag on the lower levels and flat above.

1 The building volumes are clad in a serrated (saw-tooth) perimeter brick façade, St. Joe brick (from New Orleans), commonly used on-campus. Image: Iwan Baan.

2 The brick is a mixture of warm blends and textures: a zigzag on the lower levels and flat above. Image: Joe Aker.

3 The stepping ensemble of 12 and 5-story housing and a 2-story events building with terrace, mediates between the city and the campus. Image: Joe Aker.

4 As a publicly accessible threshold, a large roof terrace mediates between living and common areas. Image: Iwan Baan.

5 The column-free dining hall is the primary social space with seating for 300. Image: Joe Aker.

5

louvered structures offer shade during hot seasons and contribute to the sustainable design. A secluded courtyard provides privacy and respite from the noise of Main Street.

The building exteriors are covered in a serrated perimeter brick façade, St. Joe brick, commonly used on campus. The bricks showcase warm blends and textures, alternating between zigzag patterns on lower levels and flat surfaces above. The facades further exhibit patterns through a combination of running bond and soldier bricks. Terracotta fins act as sunscreens and complement the masonry, particularly on the lower levels. An open-air arcade facing the quad leads to a concrete terrace, serving as a seating and observation area for quad activities.

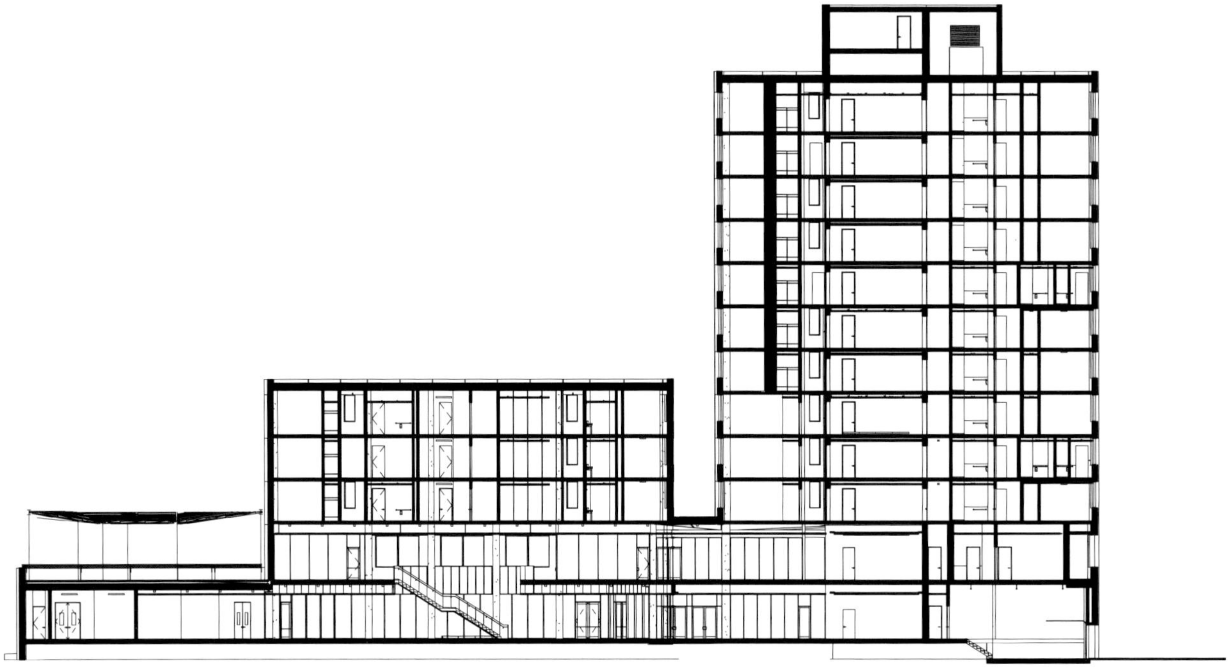

A-A long section

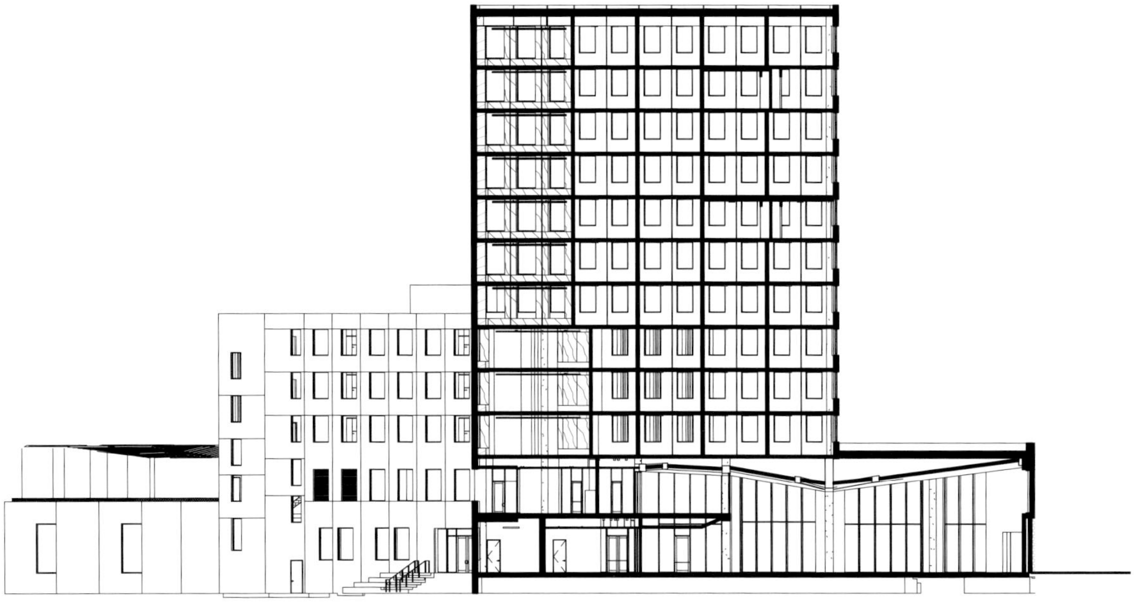

B-B short section

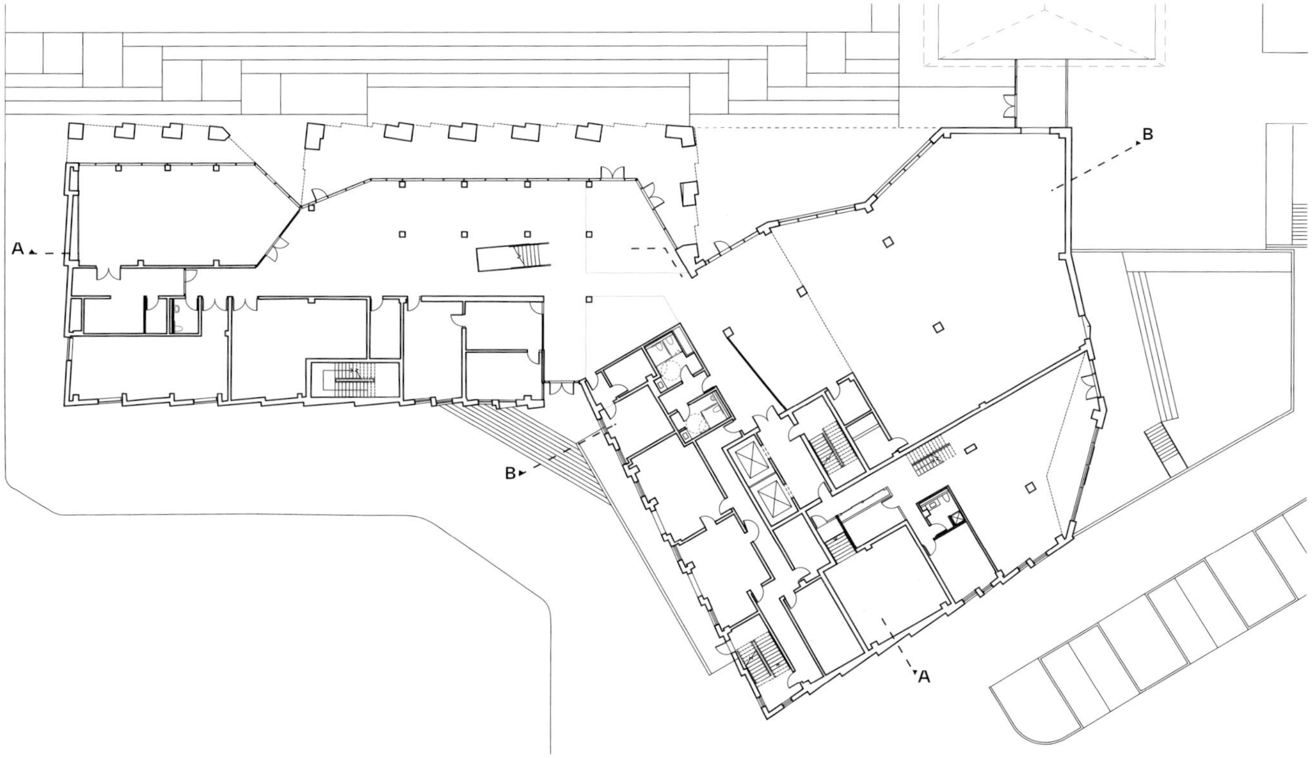

Level one

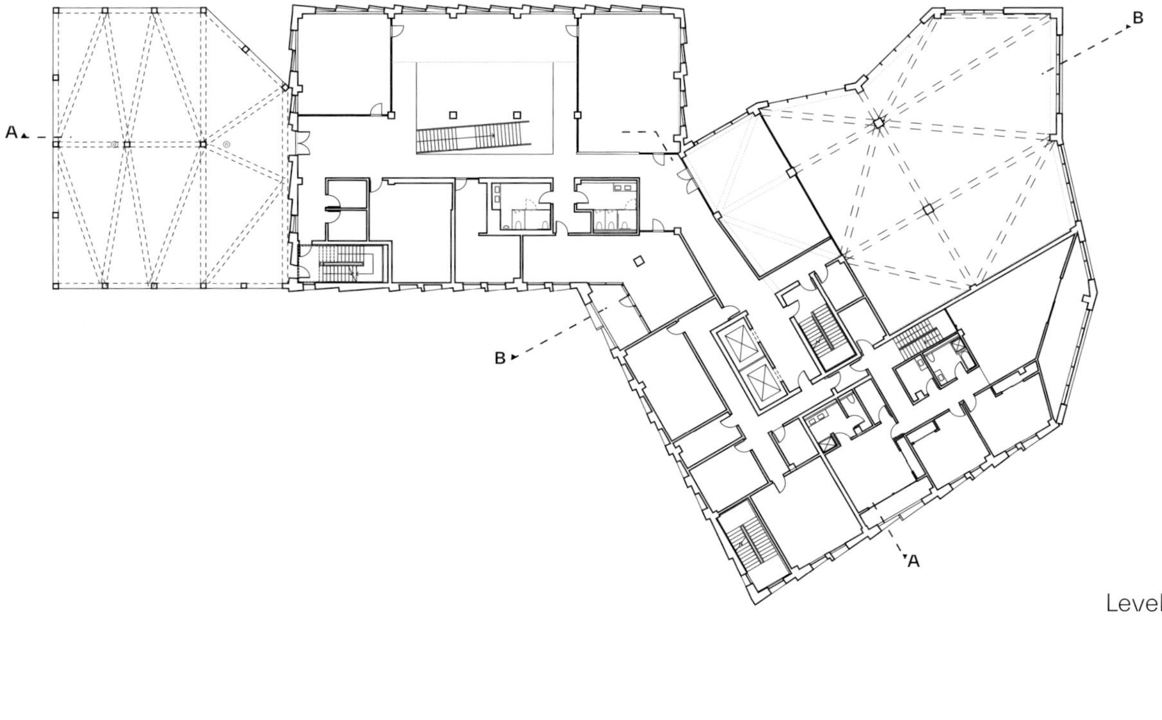

Level two

ANSYS Hall
Bohlin Cywinski Jackson

Location Pittsburgh, Pennsylvania
Function Higher Education Engineering Facility
Client Carnegie Mellon University
Completion date 2020
Architect Bohlin Cywinski Jackson
Sustainability LEED Gold Certified
Engineer Civil Langan Engineering
Engineer MEP Buro Happold Limited
Engineer Structural Barber & Hoffman
General Contractor Mosites Construction Company
Landscape Klavon Design Associates, Inc.
Photography Nic Lehoux

ANSYS Hall is a hive for teaching, prototyping, and implementation at Carnegie Mellon University. The 36,000 square foot mixed-use building was designed for the University's College of Engineering in collaboration with the engineering software company ANSYS, Inc. The goal was to create an environment that encourages learning and collaboration, while integrating gracefully with surrounding buildings.

In 2015, Carnegie Mellon University began planning an expansion to its College of Engineering: a state-of-the-art maker hub envisioned to complete a 'maker ecosystem' for engineers and the broader campus. The project started in 2016 with the aim of providing a flexible space for students to tackle real-world challenges. The design drew inspiration from architect Henry Hornbostel's early 1900s Beaux-Arts campus Mall buildings, balancing modern elements with a sense of place.

Design rationale ANSYS Hall establishes new connections between four adjacent buildings that facilitate the movement of people and materials while creating new opportunities for circulation through what had been an underutilized corner of campus. Students move easily between flexible collaboration areas, classrooms, and an expansive high-bay that opens onto a maker courtyard. An elevated bridge connects the nearby Mall with the courtyard, bringing the campus community through the site.

Naturally resilient materials were selected to gracefully weather and develop a patina over time, similar to the neighboring century-old buildings.

4

5

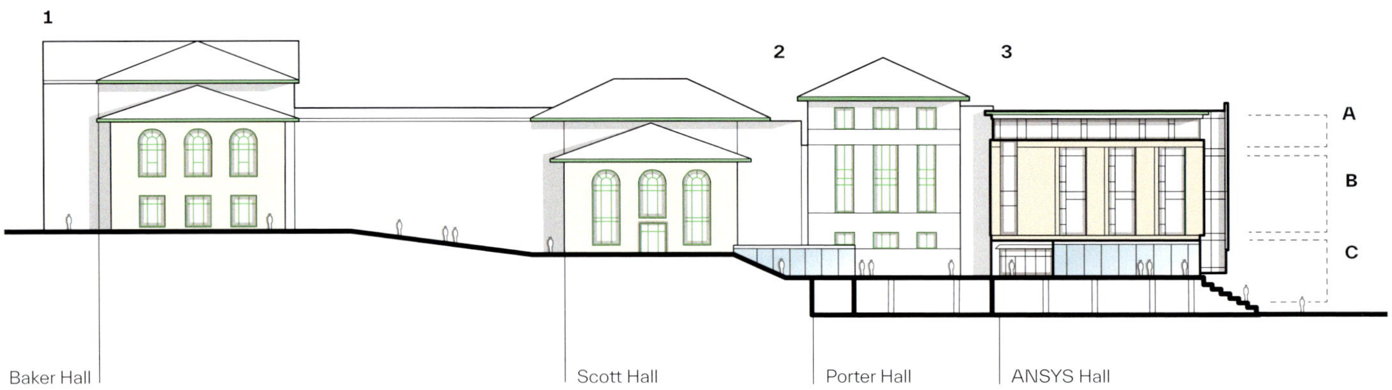

Section

1 Wing 1 (Baker Hall)
2 Wing 2 (Scott/Porter Hall)
3 Wing 3 (ANSYS Hall)

A Clerestory
B Main Floors
C Podium

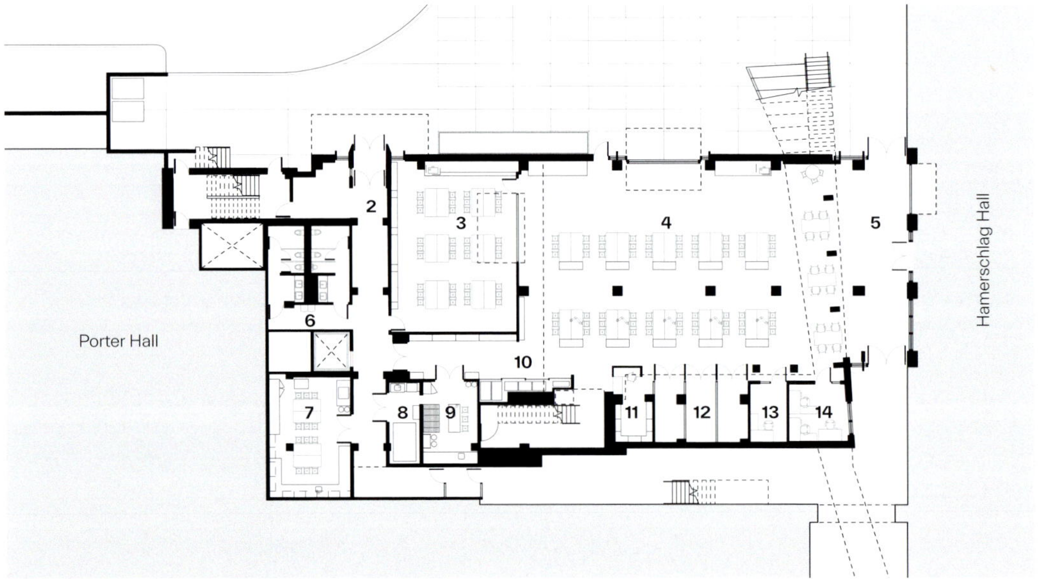

High-bay level floor plan

1 Maker courtyard
2 Main Vestibule
3 Design Studio
4 Collaboration Space
5 Hamerschlag Vestibule
6 Restrooms
7 Casting Workroom
8 Paint Workroom
9 Welding Workroom
10 3D Printer Storage
11 Tool Storage
12 Storage
13 Office
14 Shared Office

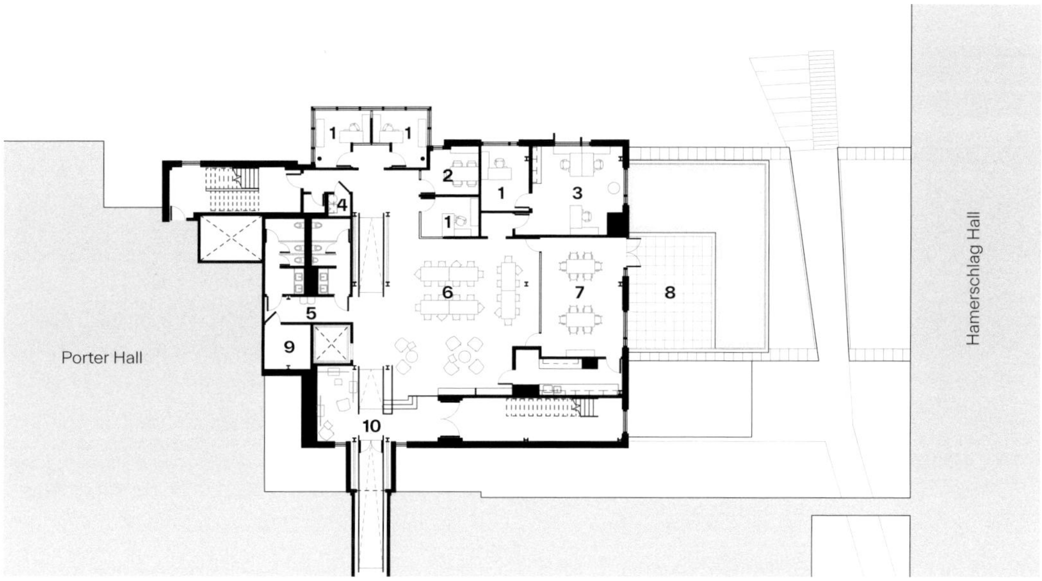

1 Office
2 Conference Room
3 Workstations
4 Custodial
5 Restrooms
6 Collaboration Space
7 Multi-purpose Room
8 Green Roof Terrace
9 IDF/ MDF
10 Collaboration Entry

Level A floor plan

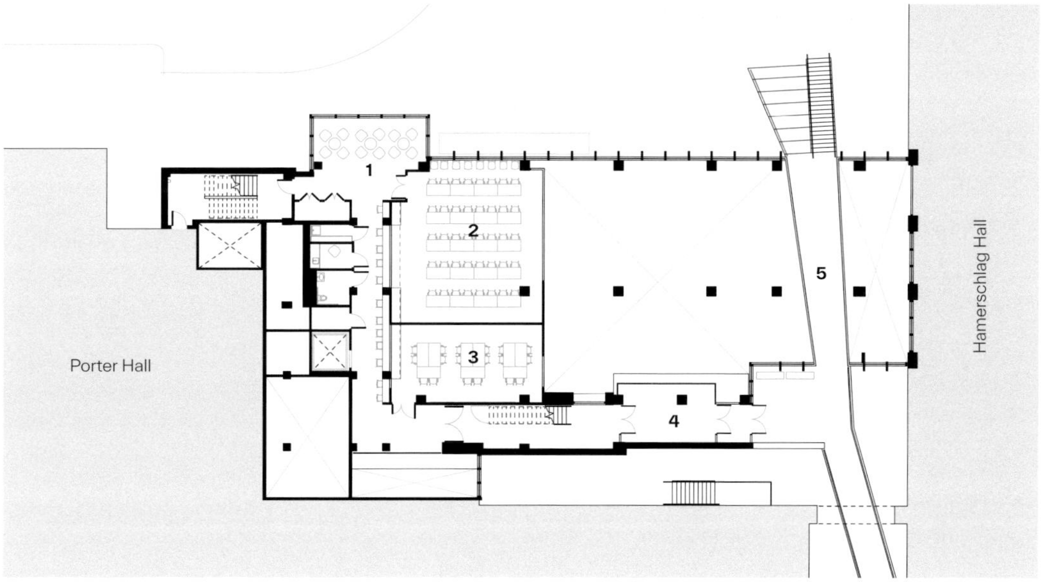

1 Collaboration space
2 ANSYS Simulation Classroom
3 Active learning
4 Vestibule
5 Connector bridge

Level B floor plan

Carnegie Mellon University
TCS HALL

1

TCS Hall
Bohlin Cywinski Jackson

Location Pittsburgh, Pennsylvania
Function Mixed-use hall
Client Carnegie Mellon University
Completion date 2020
Architect Bohlin Cywinski Jackson
Sustainability LEED Gold Certified
Engineer Structural Atlantic Engineering
MEP and Fire Protection Engineer Buro Happold Limited
Engineer Civil Langan Engineering
General Contractor Mascaro Construction
Landscape Merritt Chase
Photography Nic Lehoux

Carnegie Mellon University planned to build a new academic building on Forbes Avenue, at the western edge of its campus. The project aimed to allow for future expansion and create connections between the campus and the nearby community. It was designed to accommodate both private and university users, providing flexible spaces for collaboration while also respecting the need for privacy. The project, done in collaboration with Tata Consultancy Services (TCS), a global technology and business solutions company, resulted in an 88,000 square-foot, LEED Gold-certified facility. This building houses the Institute for Software Research (ISR), the Master of Science in Computational Finance program, the Center for Business Engagement, and the TCS Pace Port research collaboration center, all of which promote academic research, innovation, and professional development.

Contextual integration TTCS Hall is located along Forbes Avenue, a major campus gateway. This strategic placement allowed the building to blend seamlessly into both the academic and urban environments. Collaboration areas and meeting spaces were strategically located to look out over the street, bringing building occupants to the fore and creating an activated 'lantern.' Positioning TCS Hall at the edge of a sloping ravine on the site's eastern side

2

The patterned brickwork of TCS Hall drew inspiration from the original Hornbostel campus architecture, reflecting the economic significance of brick in Pittsburgh.

1 A welcoming plaza connects the project to surrounding context.

2 Patterned façade modules respond to the varied building program within.

3 TCS Hall faces onto Forbes Avenue, a primary campus gateway.

4 Positioning building services along the ravine maintained a clear architectural expression.

3

Façade diagram

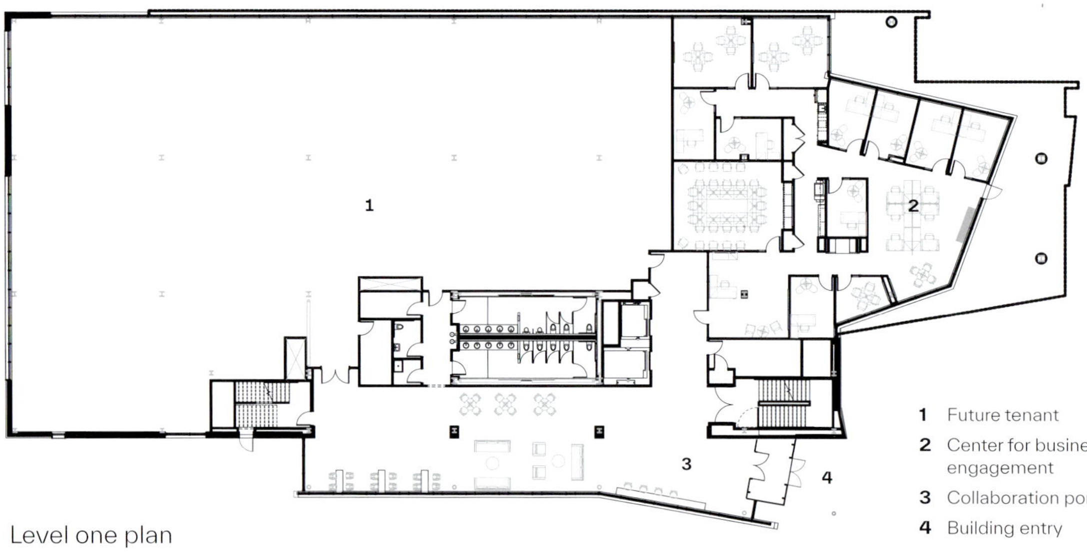

Level one plan

1 Future tenant
2 Center for business engagement
3 Collaboration porch
4 Building entry

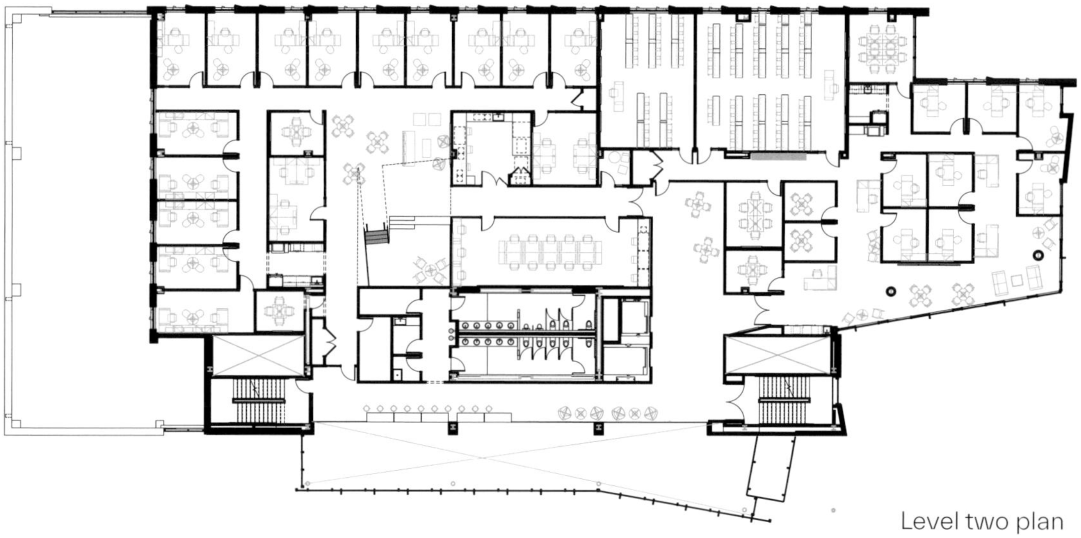

Level two plan

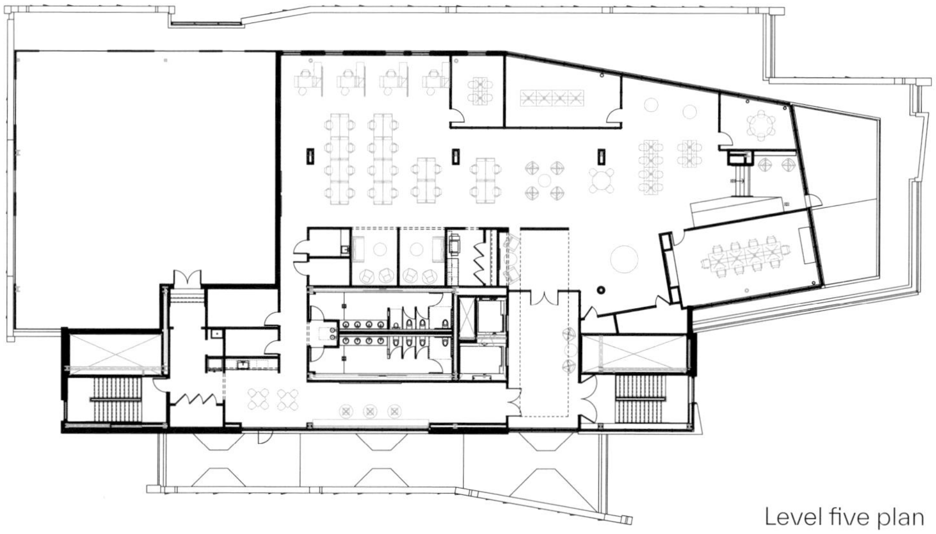

Level five plan

Zeppos Residential College

David M. Schwarz Architects

Location Nashville, Tennessee
Function Residential College
Client Vanderbilt University
Completion date 2020
Design Architect David M. Schwarz Architects
Architect of Record Hastings Architecture
Sustainability LEED Platinum Certified
General Contractor Layton Construction
Structural Engineer EMC Structural Engineers
MEP Engineer Smith, Seckman, Reid, Inc.
Civil Engineer Barge Cauthen Associates
Photography Steve Hall, Hall+Merrick+McCaugherty

Nicolas S. Zeppos College is the second of four residential college designed by David M. Schwarz Architects (DMSAS) for Vanderbilt University. The college is located at the intersection of West End Avenue and 25th Street, a short distance from downtown Nashville. Zeppos serves as a gateway for faculty and visitors entering the University and a home to 340 undergraduate students.

The building had to be both monumental and familiar. DMSAS designed the 260,000 square foot college in a collegiate gothic style that recalls the architectural heritage of schools including Oxford, Cambridge, Yale, and Princeton. The design uses limestone stone details, patterned brick, and a warm palette of interior materials to humanize the scale of the building and create a series of public and private spaces that are comfortable to residents.

A dining room, great hall, and collection of study rooms encourage interaction among the student community. In the adjoining 20-story tower, 22 long-term stay apartments are designed to attract and house visiting scholars. On the top floors of the tower are two additional convening spaces. The lower serves as a classroom and seminar space while the upper, decorated with paneled mahogany, can be used as an executive boardroom for University leadership.

1

The design uses limestone stone details, patterned brick, and a warm palette of interior materials to humanize the scale of the building.

2

Detailing work The traditional detailing including brick, carved Indiana limestone, and sandstone accents was particularly challenging as it had to work with modern cavity wall construction. The team utilized an exterior curtain wall engineer to reconcile the materials from three different suppliers into one set of coordinated shop drawings. While materials, finishes and details such as brick, Indiana Limestone and Tennessee Crab Orchard sandstone were selected for their aesthetic qualities and durability, Zeppos also attained LEED Platinum Certification.

Primary sustainability design features and strategies of the project include:

- Collecting rainwater stored and reused for landscape irrigation and public restroom toilet flushing in a portion of the project.
- Open space around the buildings and operable windows in the student rooms and apartments allow occupants to engage with nature.
- Incorporation of energy management systems and sub-metering that allow the facilities to be optimized for energy savings performance.
- Material and product selections based on life-cycle analysis, considering the environmental impact of raw material extraction, and energy and water used to manufacture products.
- Low-emitting (low VOC) material finish selections to benefit indoor air quality.
- Enhancements to indoor air quality through added air filtration and carbon dioxide level monitoring.

DMSAS has developed significant experience on a broad and diverse body of project types including civic, institutional, cultural, commercial, residential, educational, sports, healthcare, and planning projects. It believes that good design transcends typology. The Firm creates singularly unique solutions for each of its projects; a commitment that provides significant value.

3

The traditional detailing had to work with modern detailing including brick, carved Indiana limestone, and Crab Orchard sandstone accents.

4

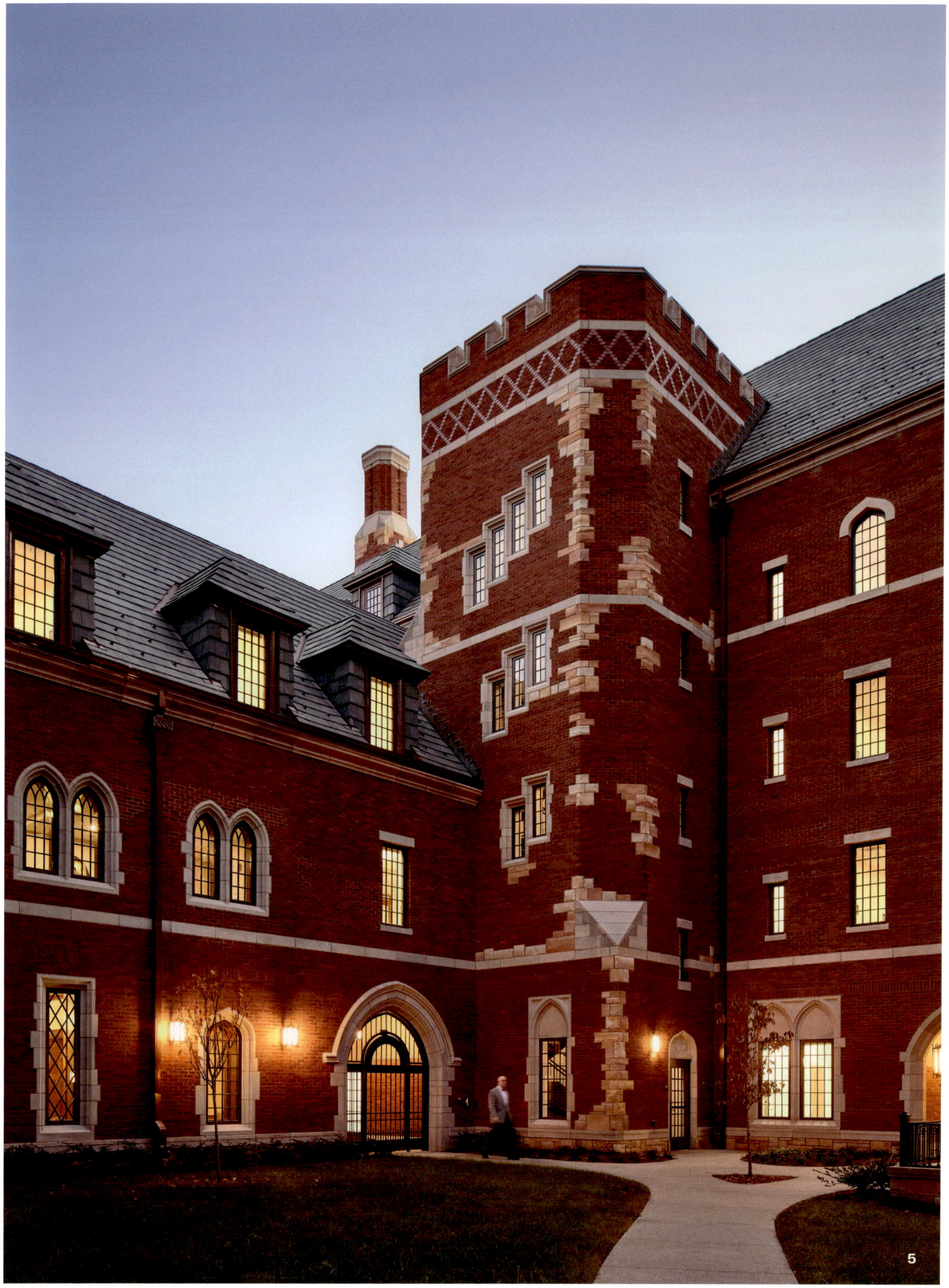

5

1 View of the southwest corner of the College featuring a large open gateway serving as the starting point for a public promenade that passes through a series of gates, portals, and courtyards to connect Zeppos with three additional colleges and the University's central Alumni Lawn.

2 Detail of covered portal entry, featuring a carved Indiana Limestone arch surround, blended field brick with decorative brick laid in a diaper pattern, Tennessee Crab Orchard stone accents, custom ornamental light fixtures and metal railings, and a custom designed ornamental gate in the distance.

3 Sited at the corner of campus at West End Avenue and 25th Street, Zeppos serves as the gateway for visitors. The building addresses the corner with a seven-story tower that marks the entrance to campus. The 310-foot West End Tower is visible in the background.

4 The Dining Room is distinguished by a fan-vaulted ceiling rendered in GFRG and acoustic plaster, accented by a large central chandelier. Bathed in light from large arched windows, the Dining Room provides capacity to seat half of the College at one time.

5 The courtyards take advantage of stair towers to add interest and character. The one visible in this image has crab orchard sandstone accents and decorative polychrome band at the top. A gated portal entrances connecting the courtyard left of the stair tower provides visual connectivity and necessary security.

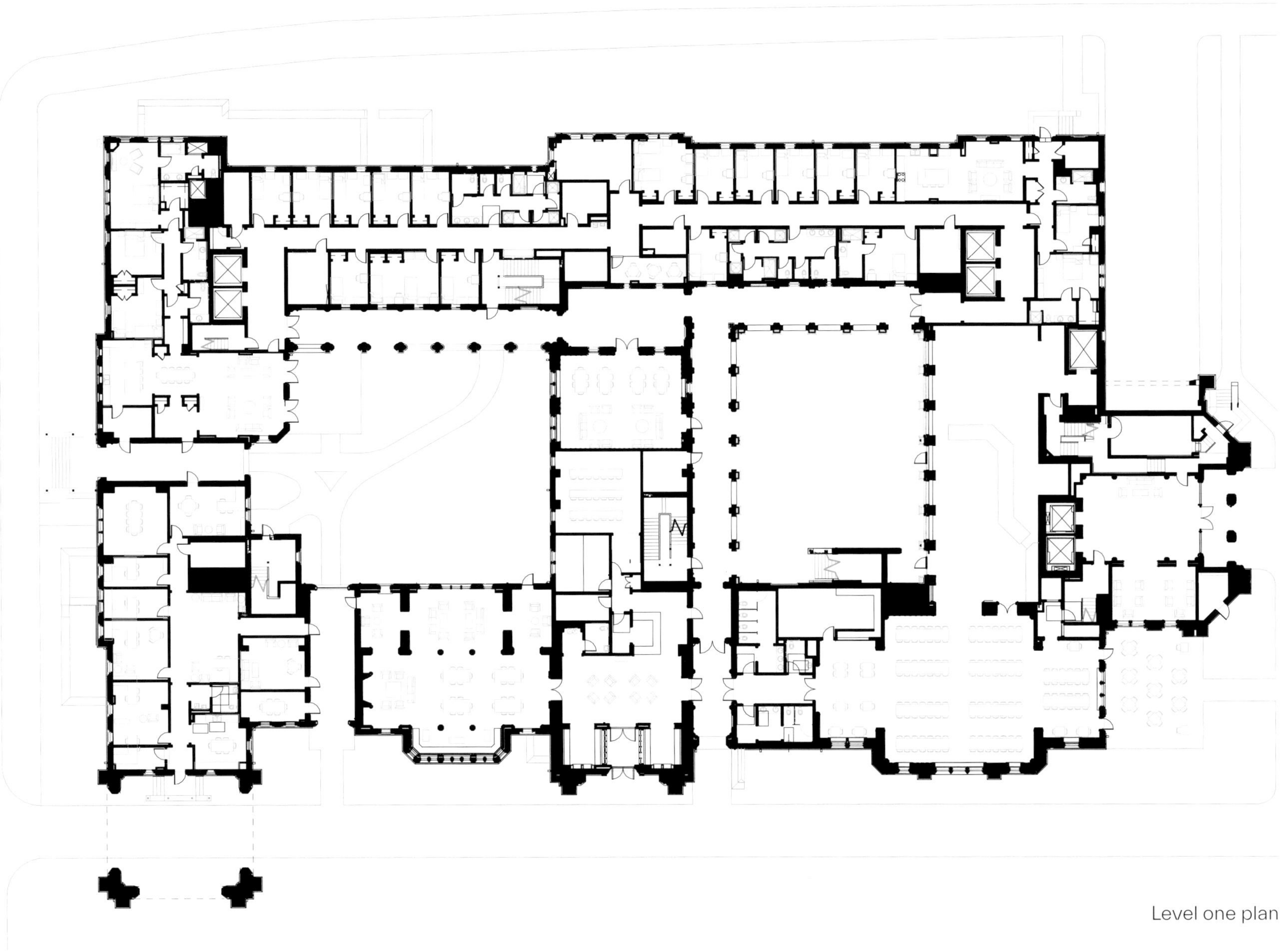

Level one plan

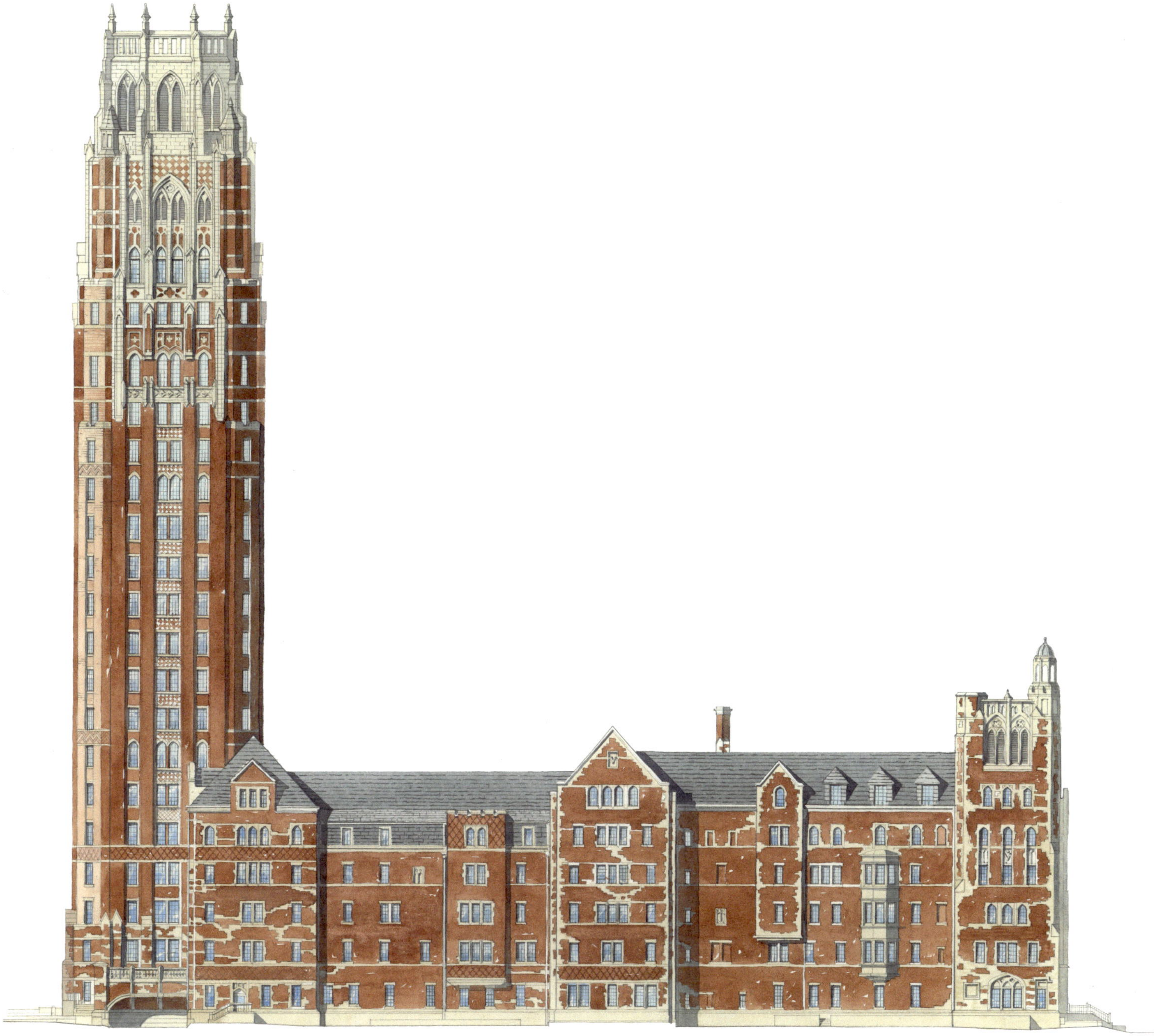

Elevation

The Owsley Brown II History Center

de Leon & Primmer Architecture Workshop

Location Louisville, Kentucky

Function Archival storage, multi-use event halls, exhibit spaces

Client The Filson Historical Society

Completion date 2017

Architect de Leon & Primmer Architecture Workshop

Landscape MKSK

Structural Engineer Tetratech

Civil Engineer BTM Engineers

Contractor Wehr / Realm

Photography de Leon & Primmer Architecture Workshop

1

The design incorporates multi-use event spaces while doubling The Filson's archival storage capacity.

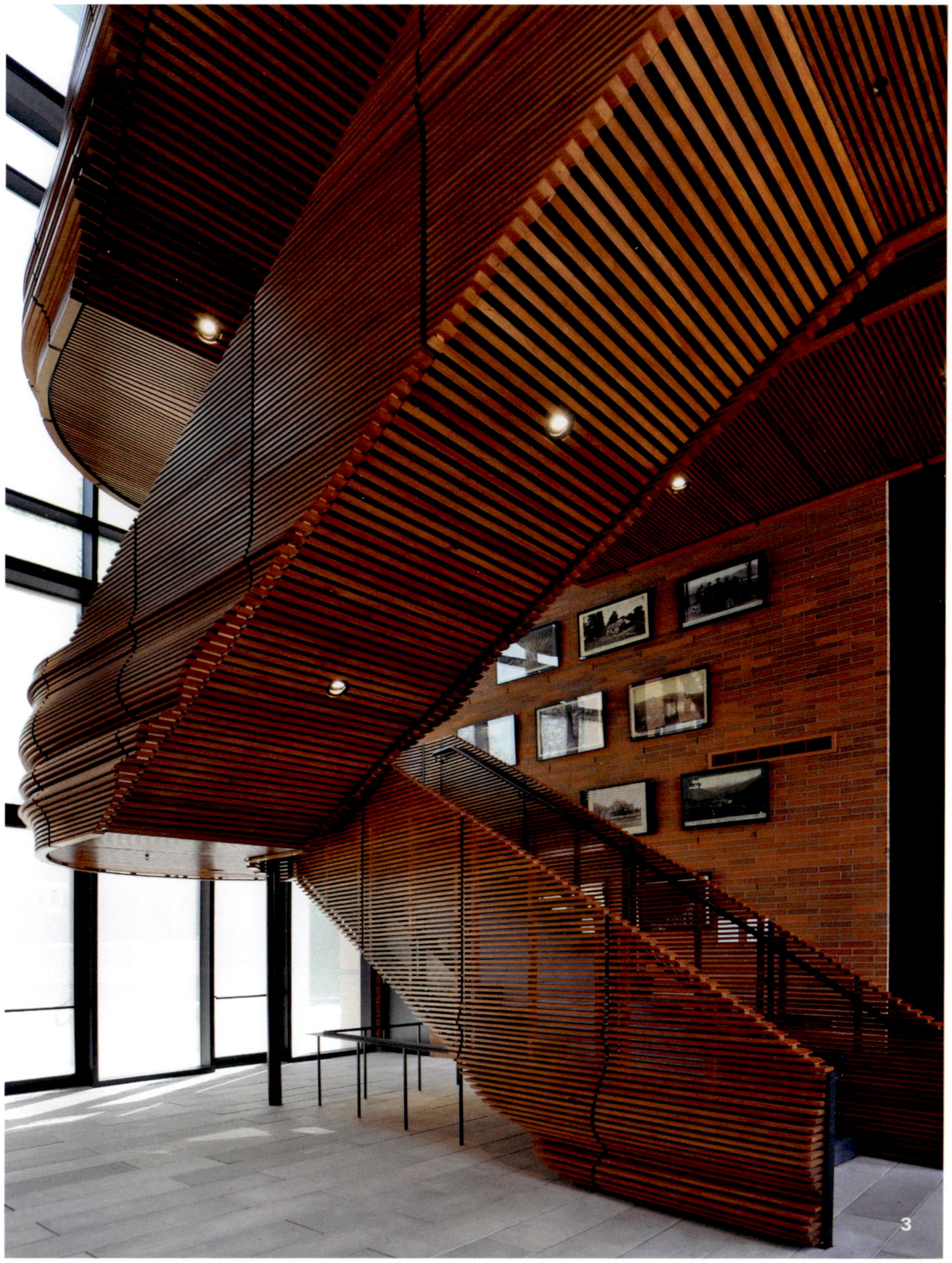

3

Based in Louisville, Kentucky, The Filson Historical Society is an organization that collects, curates, and archives the rich narrative of the Ohio River Valley region, offering an ambitious range of educational programs and cultural resources that support this focus. Headquartered within the historic neighborhood of Old Louisville, The Filson is encompassed by the nation's largest contiguous collection of Victorian mansions, developed during the Great Southern Exposition in the late nineteenth century. After 130 years of collection and operation, The Filson outgrew its existing home in the Ferguson Mansion, an historic example of Beaux Arts architecture.

The Owsley Brown II History Center is a new 30,000 square foot structure—one component of a unified campus expansion that includes a renovated Ferguson Mansion, an adjacent Carriage House, and a new public plaza. By visually revealing its internal functions to the community and emphasizing the use of layered contemporary construction techniques, the new History Center presents a deliberate spatial contrast to the Ferguson Mansion, with its dense masonry walls and small apertures. Programmatically, the design incorporates multi-use event spaces while doubling The Filson's archival storage capacity.

Guided by extensive, contextual research of the Old Louisville preservation district, and inspired by the architectural components of the Ferguson Mansion, the new building recalls traditional period details through wood—specifically the exclusive use of stained poplar (also known as the 'tulip tree'—the state tree of Kentucky). Ornate ceilings, elaborate staircases, and other elements are reinterpreted with contemporary fabrication techniques that investigate proportionality, depth, and layering of materials. In an abstract reference to The Filson's historical focus on the Ohio River Valley, the wood-lined interiors of the primary public spaces allude to conceptual precedents such as water droplets, ripples, and refracted surfaces.

Elements are reinterpreted with contemporary fabrication techniques that investigate proportionality, depth, and layering.

As a testament to its inclusive design process, The Owsley Brown II History Center is the only large-scale building within historic Old Louisville to receive unanimous approval by the city's Landmark Commission in over 30 years. Following an eight-year period of design, private fundraising, and construction, the project was completed in March 2017.

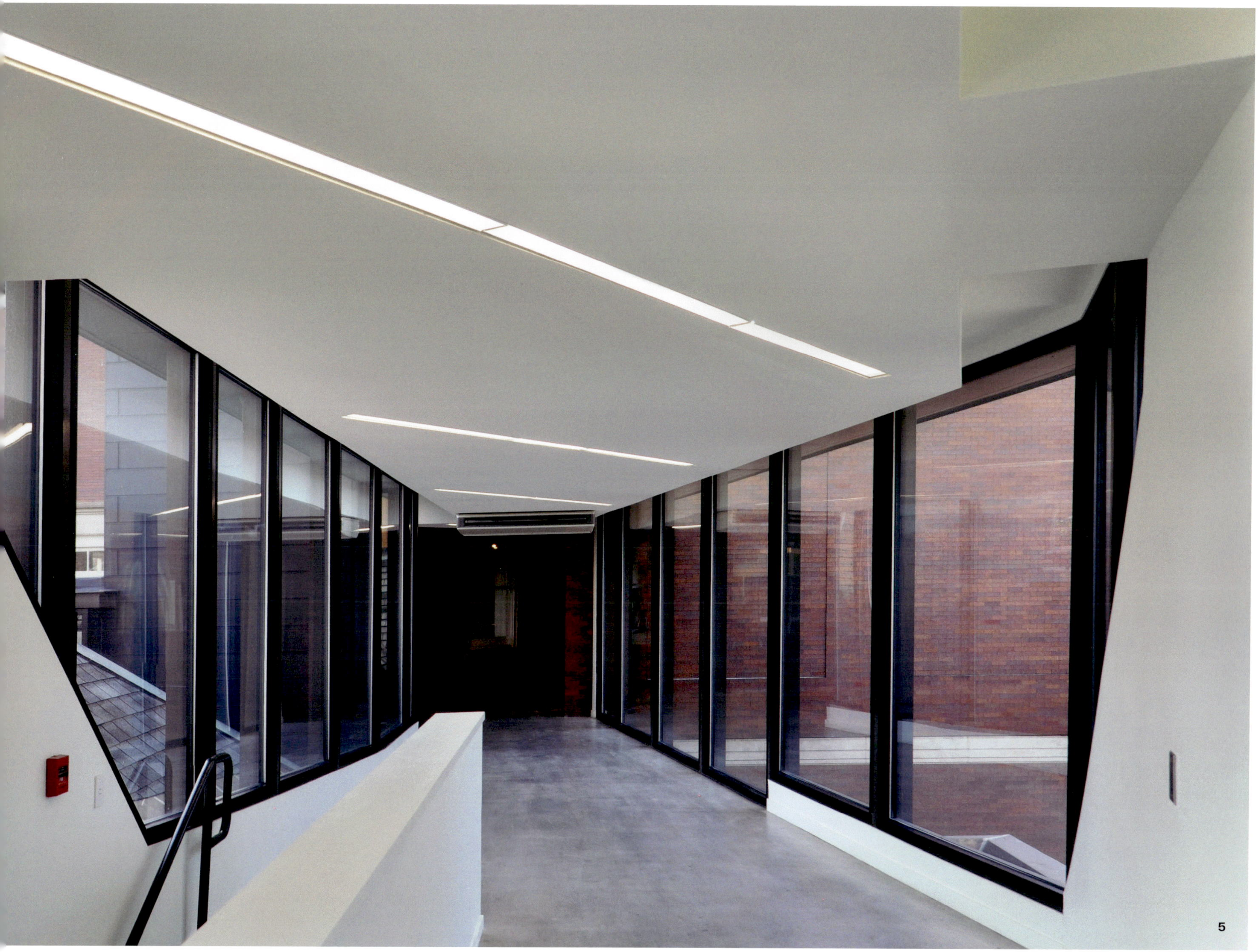

5

1 View from northwest.

2 View from southeast with pedestrian bridge.

3 View of grand staircase.

4 Detail view of brick façade.

5 Interior view of pedestrian bridge.

Site section

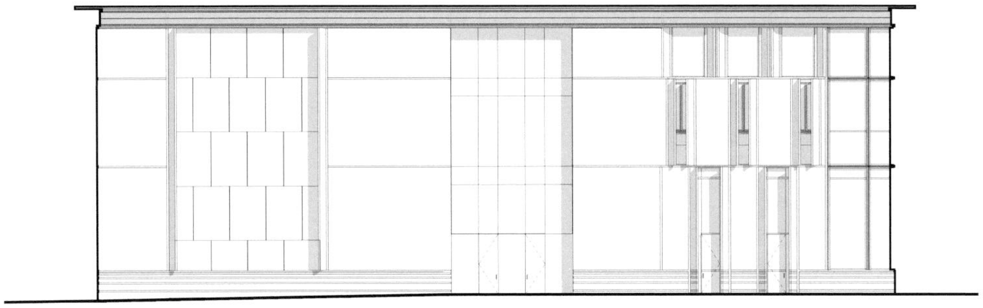

North elevation

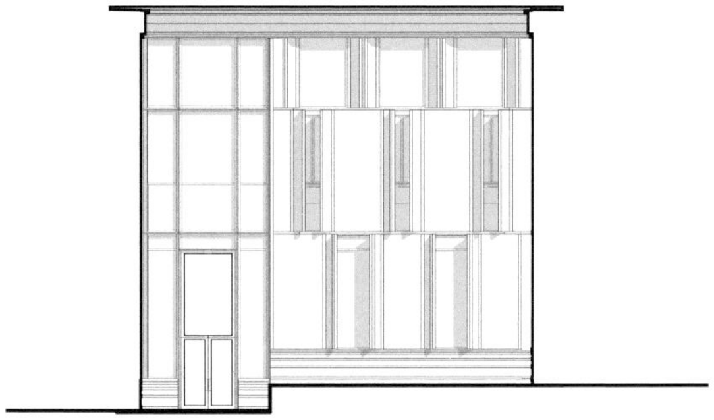

West elevation

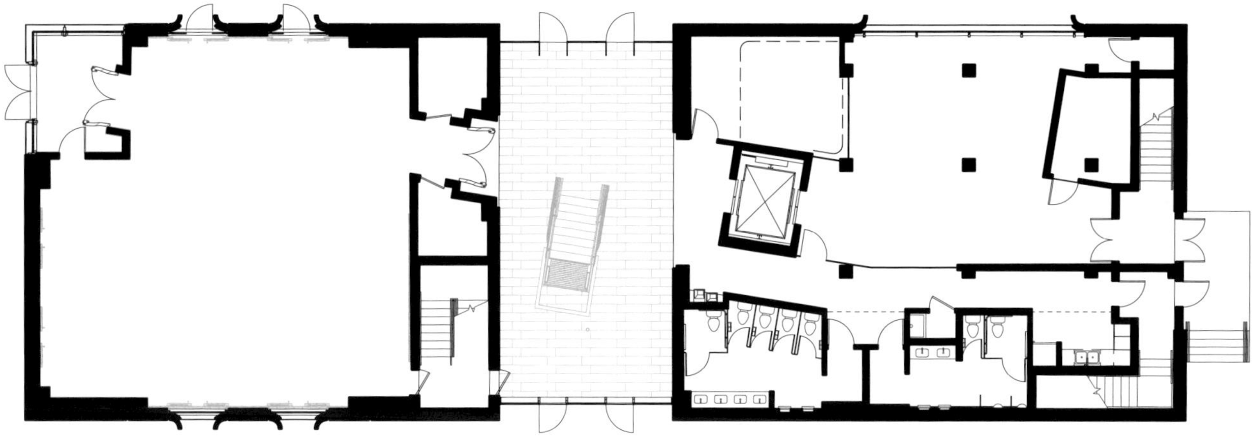

Ground floor plan

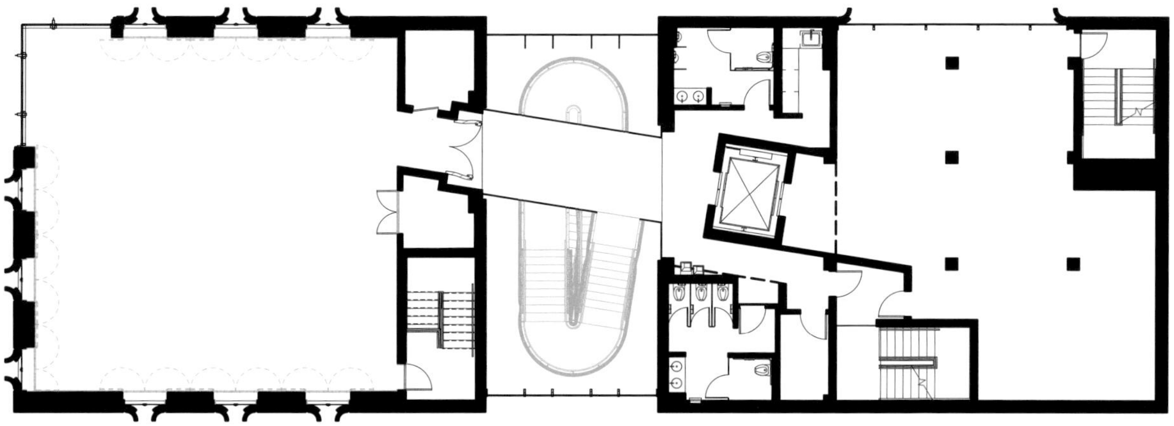

Terrace floor plan

Biblioteca Latinoamericana

Ehrlich Yanai Rhee Chaney Architects

Location San Jose, California
Function Public library
Client Redevelopment Agency of San Jose
Completion date 1999
Architect Steven Ehrlich Architects
Landscape Patricia O'Brien Landscape Architecture
Architect of Record Garcia Teague Architecture
Contractor Alvarado Construction
Photographer Tom Bonner

Situated just a mile south of downtown San Jose, this branch of the San Jose Public Library was designed to meet the needs and interests of the local community. Specifically, it caters to a primarily Hispanic population. The library, known as Biblioteca, was established in 1974 by a group of mothers from a local Catholic church. Their goal was to promote bilingual education and the study of Latino culture within their immigrant community. In 1999, Biblioteca moved to its current location, supported by grants and private funding.

Biblioteca Latinoamericana is now home to one of California's largest collections of Latin American literature and history. It's a cherished landmark in San Jose and offers a variety of year-round programs for families to enhance learning and literacy. Beyond books, it also serves as a vibrant community hub. It provides amenities like free Wi-Fi, a computer lab, art exhibitions, counseling services, and social and recreational activities. The space includes a Youth Center gym where teenagers can play basketball, an area for mothers to engage in activities with their young children, and during the pandemic, it also offered free vaccine distribution.

The design of Biblioteca Latinoamericana was crafted by Steven Ehrlich Architects, now known as Ehrlich Yanai Rhee Chaney Architects. The focus was on fostering social

1

The entrance features a large brick, stone, and masonry wall, which blends traditional Latin-American motifs with modern patterns.

2

and intellectual interactions. The library, spanning 15,000 square feet, is connected to the adjacent 17,000 square-foot Washington United Youth Center through a landscaped plaza.

Community-centric design The location is a flat, densely populated residential area. The architects aimed to honor the Latino heritage of the community through the library's front entrance. The entrance features a large brick, stone, and masonry wall facing First Street, which blends traditional Latin-American motifs with modern patterns. The choice of structural brick was inspired by ancient sites in Oaxaca, with its versatility in color and texture. Illuminated at night, this design symbolizes the merging of history and modernity.

Both the Library and Youth Center share a material palette featuring varying shades of yellow, brown, and gray brick. Glass and steel elements add lightness and transparency. The library's front façade includes a glass-enclosed seating area that visually connects patrons inside with the city outside. Both buildings have central spaces defined by concrete block and brick walls, with metal roofs and clerestory windows. Glass doors connect both structures to the outdoor plaza. The Youth Center, located next to an elementary school, encourages sports participation with its gymnasium and athletic equipment.

The courtyard features two trellised pathways that welcome pedestrians. A fountain by artist Elsa Flores adds a soothing ambience, and a permanent stage hosts events like celebrations, weddings, and musical performances.

Benefits of brick Brick was chosen not only for its aesthetics but also for its practical advantages. It's a natural, sustainable material made from clay. Brick maintains its colors over time, reducing the need for repainting. Its durability, low maintenance, and energy efficiency make it cost-effective in the long run.

The choice of structural brick was inspired by ancient sites in Oaxaca, with its versatility in color and texture.

4

1 Detail of Biblioteca Latinoamericana's large brick, stone, and masonry entrance wall showing the library's front glass-enclosed seating area.

2 The Illuminated front entrance wall facing First Street—symbolizing the merging of history and modernity.

3 Brickwork blending traditional Latin-American motifs with modern patterns.

4 Glass and steel elements add lightness and transparency.

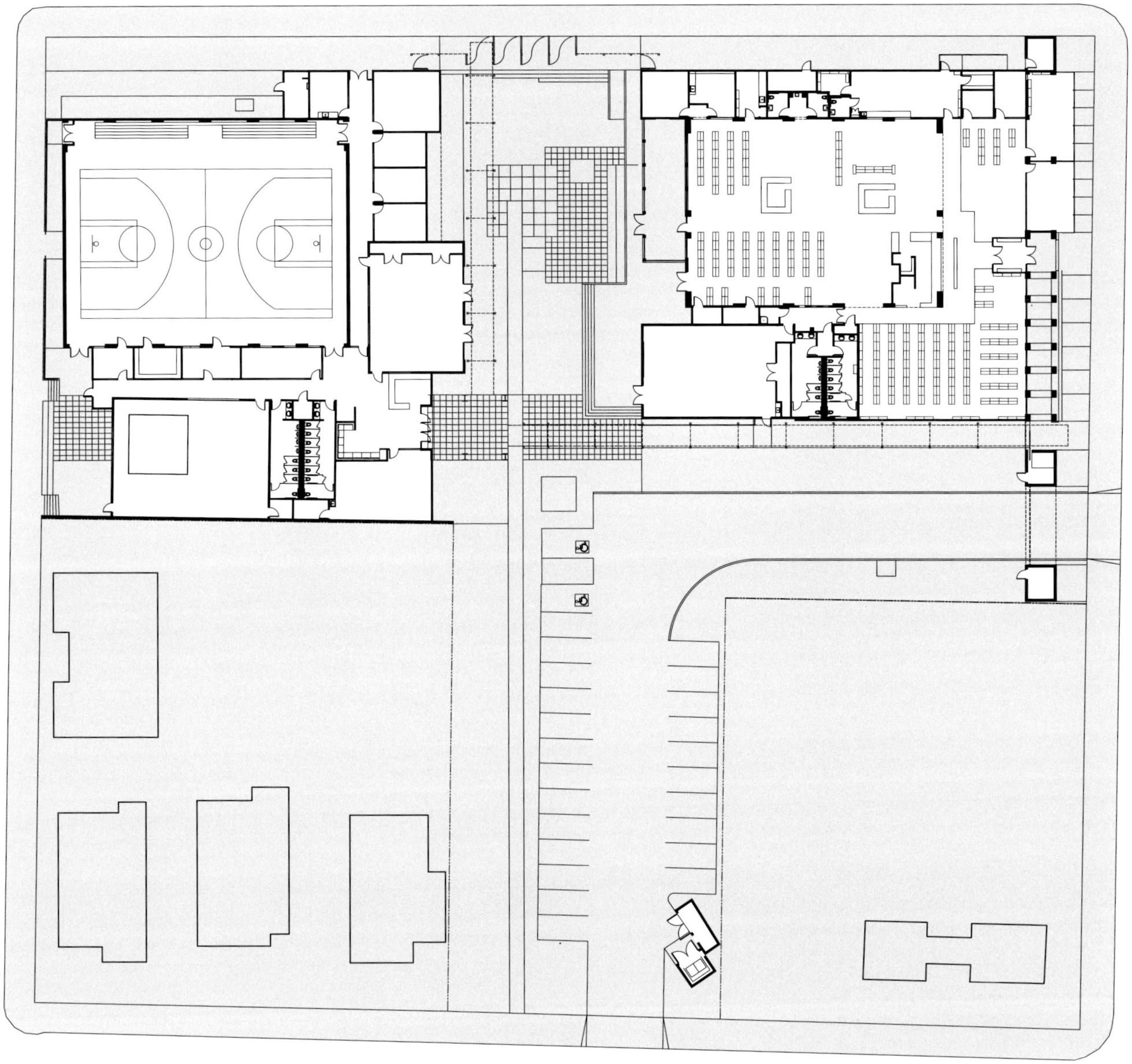

Ground floor plan

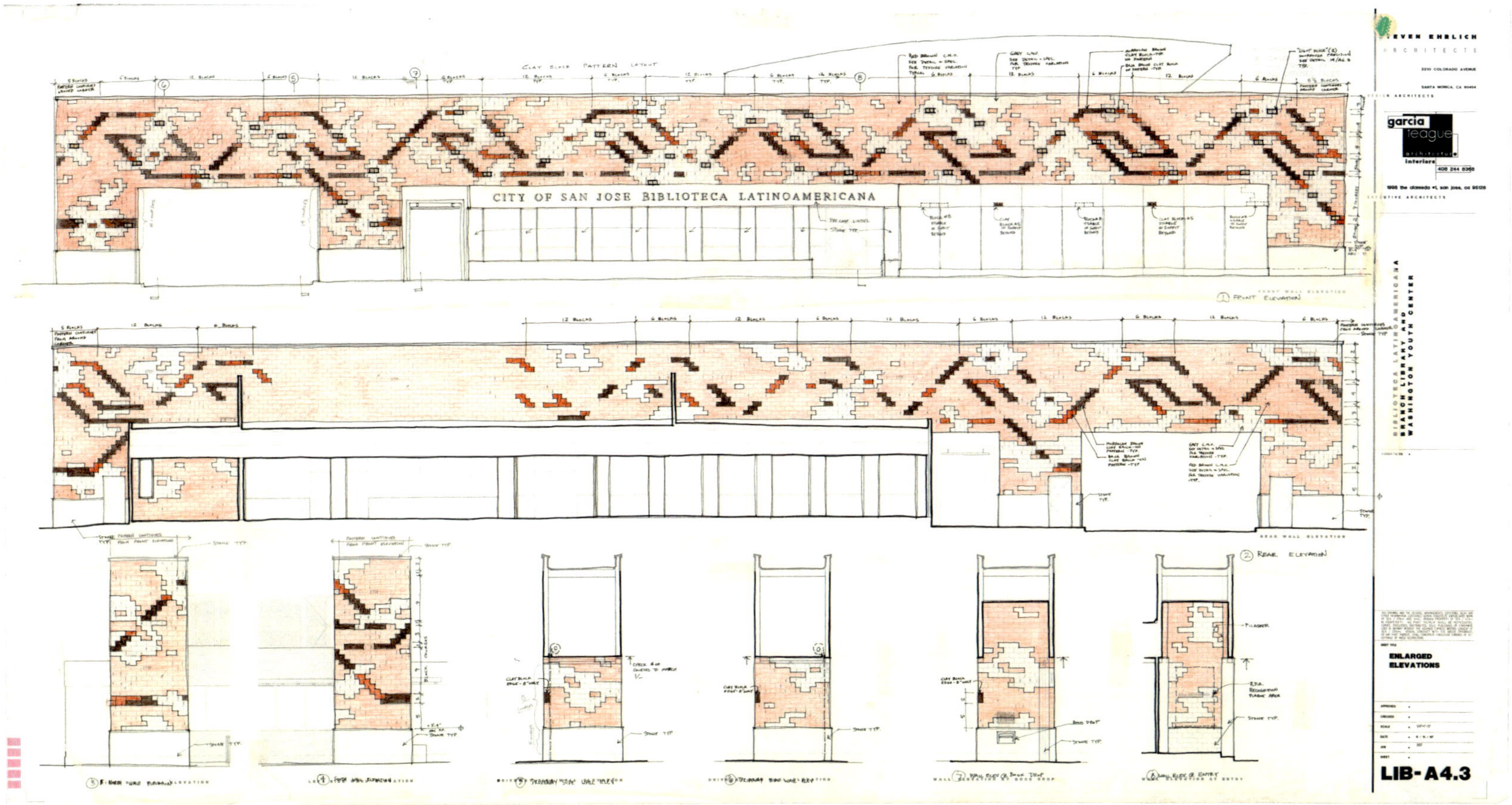

Brick plan

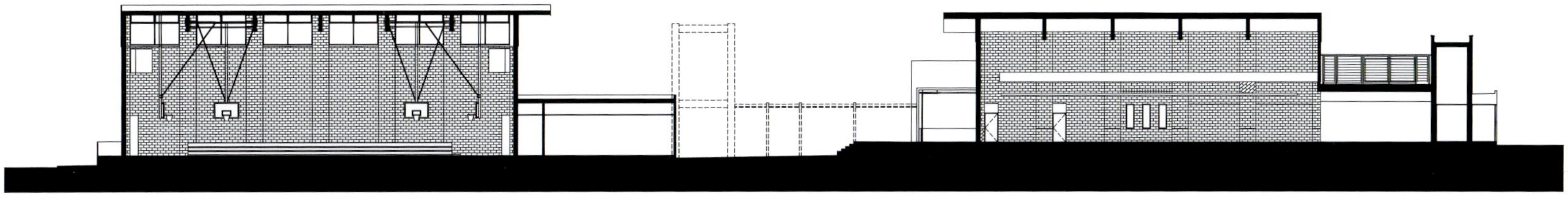

Section

DINEEN HALL

1

Dineen Hall
Gluckman Tang Architects

Location Syracuse, New York
Function Law School
Client Syracuse University College of Law
Completion date 2014
Architect Gluckman Tang Architects
Sustainability LEED Gold certified
Landscape CLEAR Landscape and Urban Design
Engineer Structural John P Stopen Engineering Partnership
Engineer Mechanical M/E Engineering
Engineer Civil GHD
Photography Ty Cole/OTTO

The 5-story building breaks down in scale by incorporating volumes featuring different colors and textures of brick.

The Syracuse University College of Law wanted a building that would meet the various needs of a modern law school in a clear and easy-to-understand way. They asked for the Dineen Hall design to include elements like size, scale, proportions, textures, and light. The building is five stories tall, covered in bricks, and has received a LEED Gold rating. It offers more than 200,000 square feet of space, fulfilling all the requirements for the School of Law, including spaces like a law library, classrooms, offices, and a ceremonial courtroom.

Dineen Hall is located in Syracuse University's new West Campus. This building stands out among other large brick structures in the area. We aimed to make it fit in better with the existing architecture of the main quad. To achieve this, we used big, overhanging architectural structures covered in differently textured and colored bricks. A new park, 60 feet wide, separates the building from a nearby parking garage and provides a nice path for students to the west campus.

Façade design and brick use The building's design features large blocks of brickwork raised above a sloping site. These blocks hover over transparent public spaces enclosed in glass: the library, a large collaborative area, a courtroom/auditorium, and a central celebratory space. This public layer is covered by a fragmented curtain wall, which resembles a virtual grove and lets in light from different directions.

To break down the building's size and bring in daylight, we divided it into smaller blocks. These blocks are formed into 3D L-shaped sections that interlock, with some extending beyond a background of lighter-toned bricks. This interplay creates spaces with varying degrees of light and shadow. Different shades and textures of bricks distinguish these blocks. Inside each smaller volume, we used vertical windows for classrooms and horizontal windows for administrative and faculty offices.

Making sure all public areas in the building have natural light was a major goal. To achieve this, we created a raised central courtyard with a green roof, skylight, and glass-

3

4

covered atrium. The skylight and atrium provide light to the lower floors, while the green roof offers an outdoor space and a view from the upper floors. The sky-lit central atrium connects essential parts of the building and introduces natural light throughout. It also features a distinctive concrete-and-steel staircase.

At the heart of the building is a teaching courtroom and auditorium. This space is designed for mock trials and observing actual court proceedings. It seats 200 people and is marked by a white glass section interrupted by large windows, showcasing the law school's accessibility and openness. A prominent staircase beside the courtroom allows passersby to glimpse the proceedings and connects to the lower level, which opens to the west.

1 East façade and main entrance

2 A sloped window frames a view through the ceremonial courtroom.

3 A white-glass volume expresses the volume of the ceremonial courtroom to passers-by.

4 The Law Library's wood-clad rare books room illuminates the area at night.

5 A sky-lit atrium brings light deep into the building and frames an architectural concrete and steel stair.

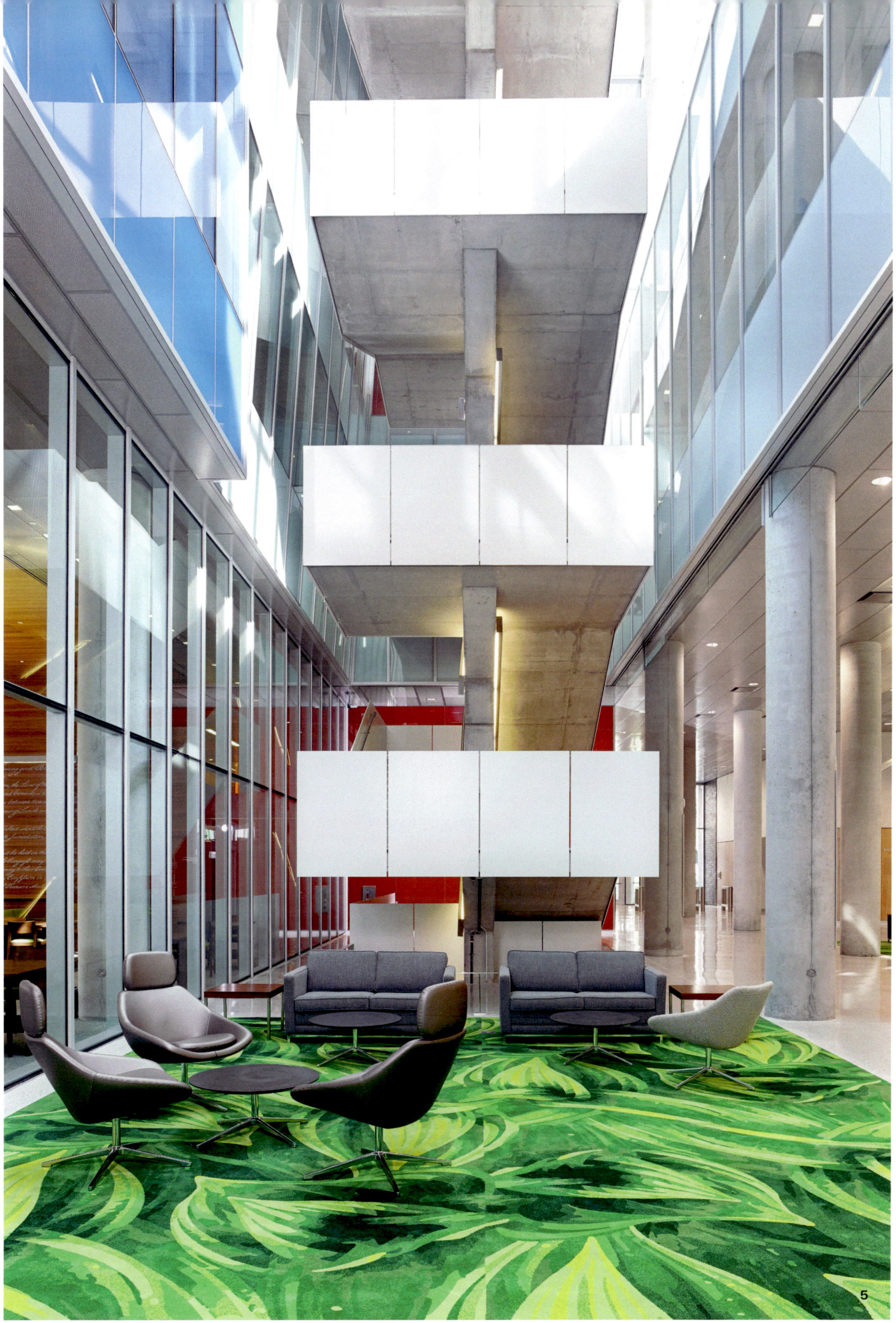

The sky-lit central atrium connects essential parts of the building and introduces natural light throughout.

Section

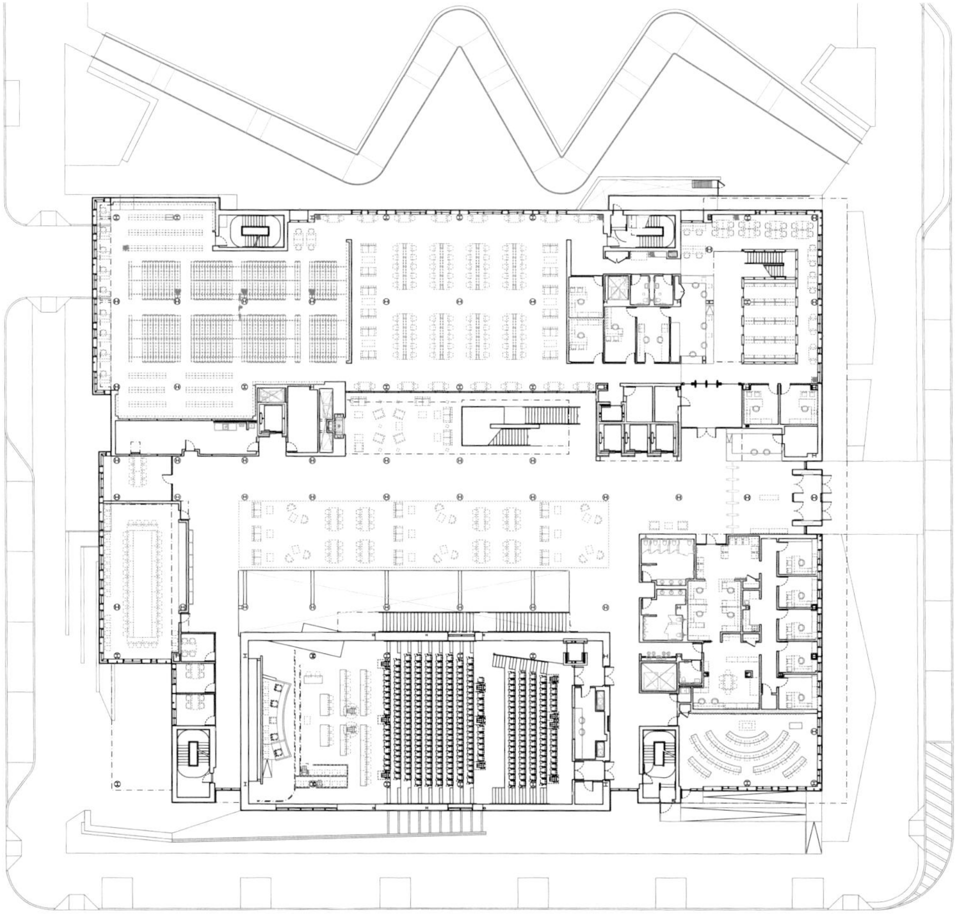

Level one plan

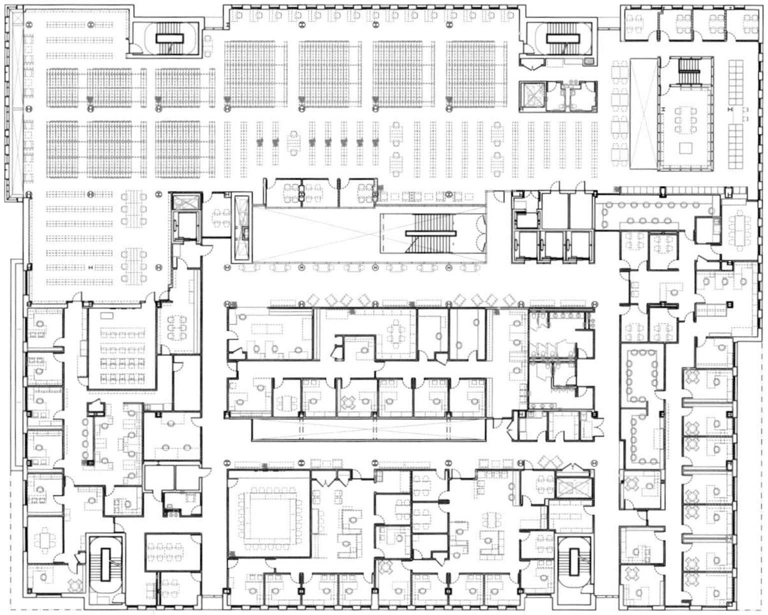

Level three plan

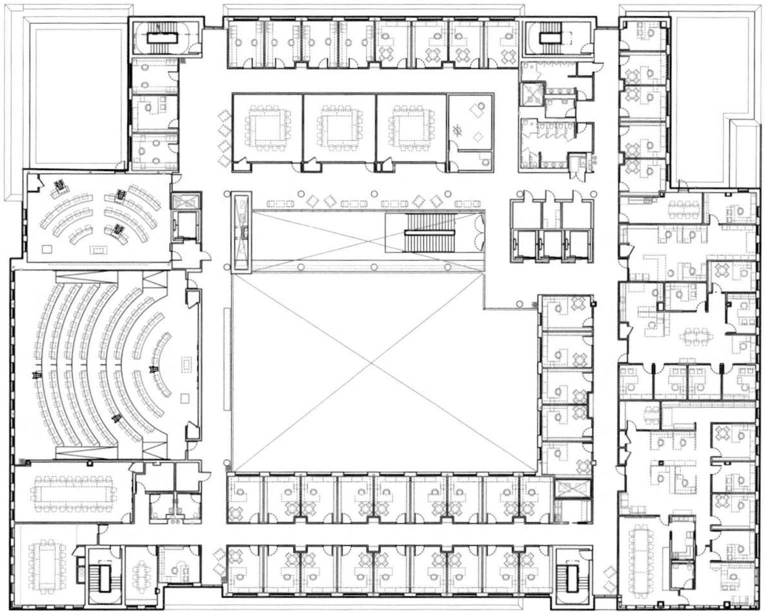

Level five plan

The Iribe Center for Computer Science and Engineering

HDR

Location College Park, Maryland

Function Centre for Computer Science and Engineering

Client University of Maryland

Completion date 2019

Architect HDR

Landscape Mahan Rykiel Associates

Structural Engineer Hope Furrer Associates Inc.

Contractor Whiting-Turner

Photography Dan Schwalm, John Consoli, Nick Seitz

Funded in part by Brendan Iribe, this gateway building connects the University's new Innovation Campus to the rest of campus, to the city, and beyond. Through virtual reality technologies, a person sitting in Vienna will be able to operate a drone from inside the Iribe Center, while a graduate student in Palo Alto looks on. But it doesn't only engage people inward. It transmits data out—connecting to the academic community across the humanities, engineering and hard sciences—to embrace industry partnerships.

The building is comprised of two main components—a six-story instructional and research space, and a 300-person auditorium—joined by connector space, and blurs the line between inside and outside. Hacker spaces allow students, faculty, and industrial and community partners to share ideas through workshops, presentations and lectures, and to work on projects individually or in teams. Open work areas and makerspaces filled with natural light, flexible classrooms, and team spaces foster active learning, and welcoming common areas extend the collaborative environment to the broader community.

1

One of the challenges of designing the Brendan Iribe Center for Computer Science and Engineering was projecting an image of technological advancement on the University of Maryland's flagship College Park campus, which is deeply rooted in a classical neo-Georgian architectural tradition. To achieve this balance, the design team deconstructed the classic design, removing the Maryland style brick façade that characterizes the other buildings on campus and leaving the highly ordered structural grid beneath. This approach effectively 'peeled' the skin off the building to reveal the buzz of activity inside.

Materiality A Glen-Gery brick is brought back as an anchor, grounding the building as the main exterior feature of the Antonov Auditorium. The brick ties in the very modern building with the historic brick knee walls fronting the road and the brick archways of the founders' gate at the entrance to campus. Contemporary design technologies allow for a non-traditional rethinking of brick, a conventional material that is important to campus heritage. Using parametric modeling, an abstract pattern was translated into a system of bricks for the auditorium wall inside the building.

In most buildings on campus, brick is portrayed as structural but in reality it is only a veneer. The application of brick in the Iribe center was manipulated in such a way that exposes its non-structural function. The brick pattern along the interior walls makes the brick appear to be moving, in what is essentially a sculptural application of brick. This was intended to be a more authentic and sincere or 'real' representation of the how the brick functions. While the pattern signifies movement and information flow, the innovative design techniques utilized to achieve it emphasize ideas such as virtual reality and technological innovation.

A kinetic hub Fusing the physical reality of a bricks and mortar building with advanced virtual reality technologies, the new center is envisioned as a kinetic hub for computer sciences on campus and beyond. A singular design, it is easily visible from its location on a prominent intersection. The façade, characterized by an inventive curtain wall system that controls solar gain while creating an optical illusion of movement, expresses the dynamism of the science conducted inside. Inherently flexible, it is designed to morph with inevitable changes in computer sciences, while also creating a technology-infused and memorable image.

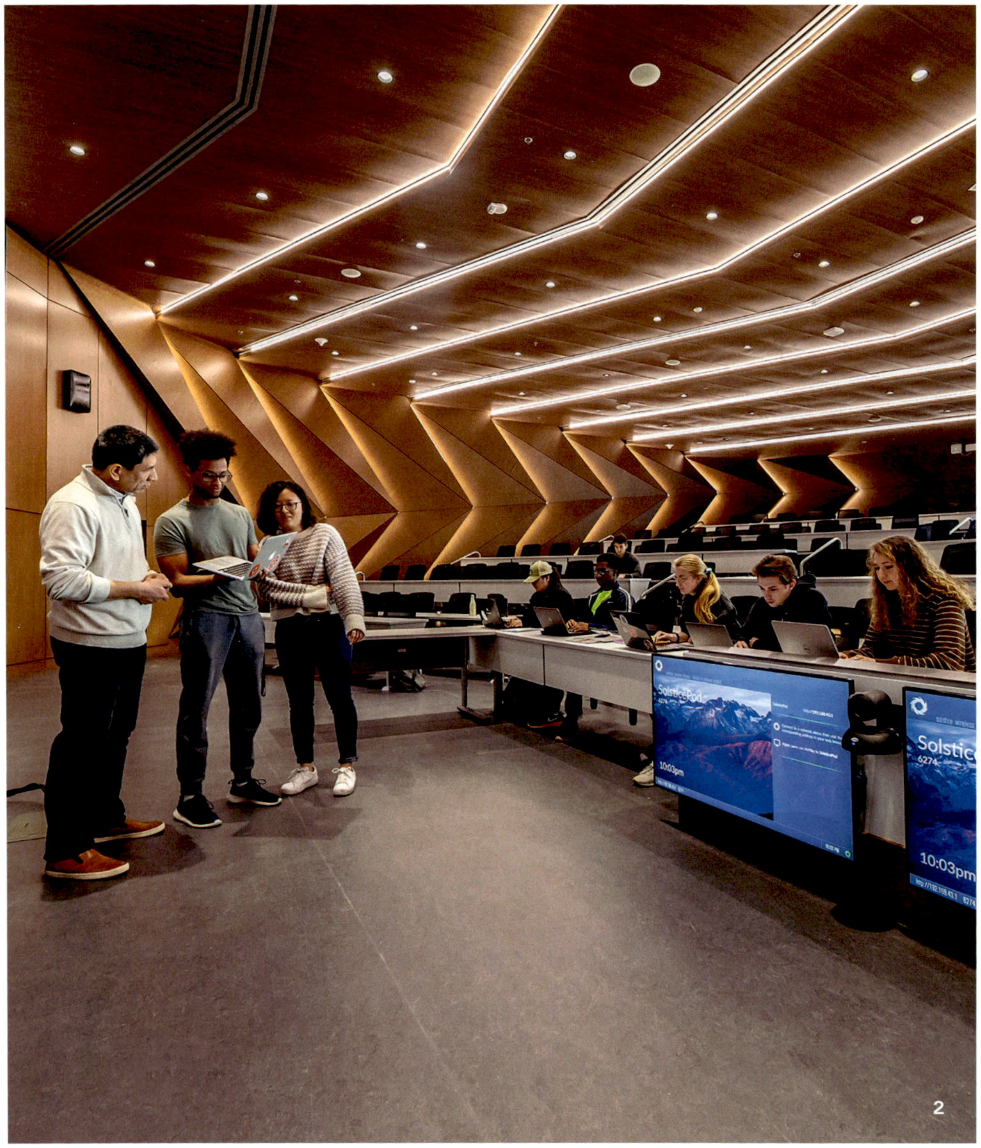

2

Using parametric modeling, an abstract pattern was translated into a system of bricks for the auditorium wall inside the building.

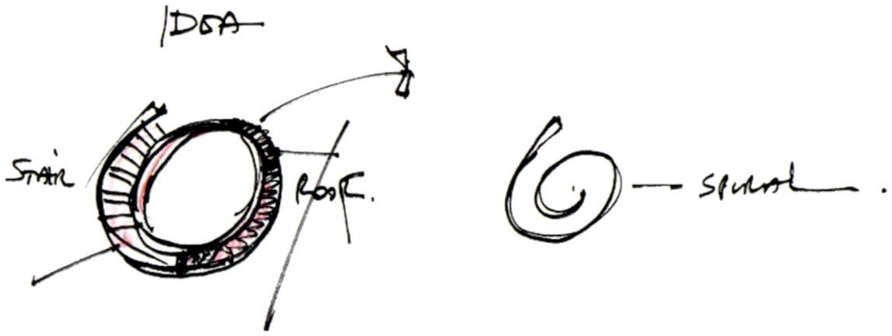

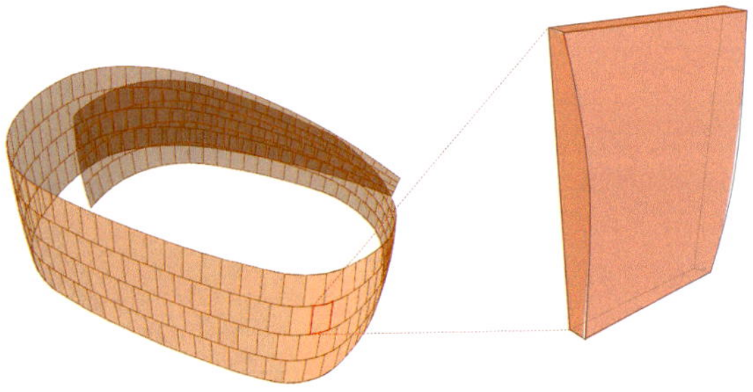

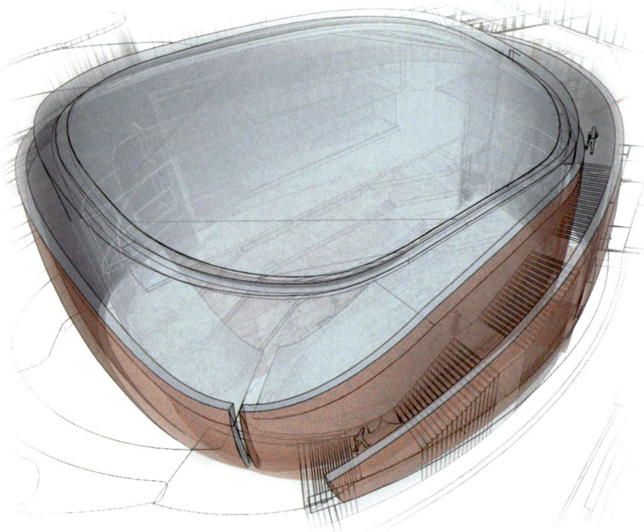

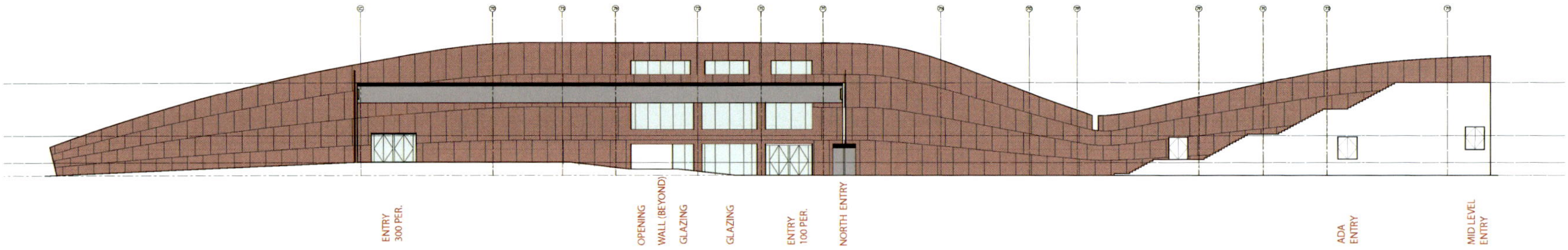

Concept sketch, renders and elevation

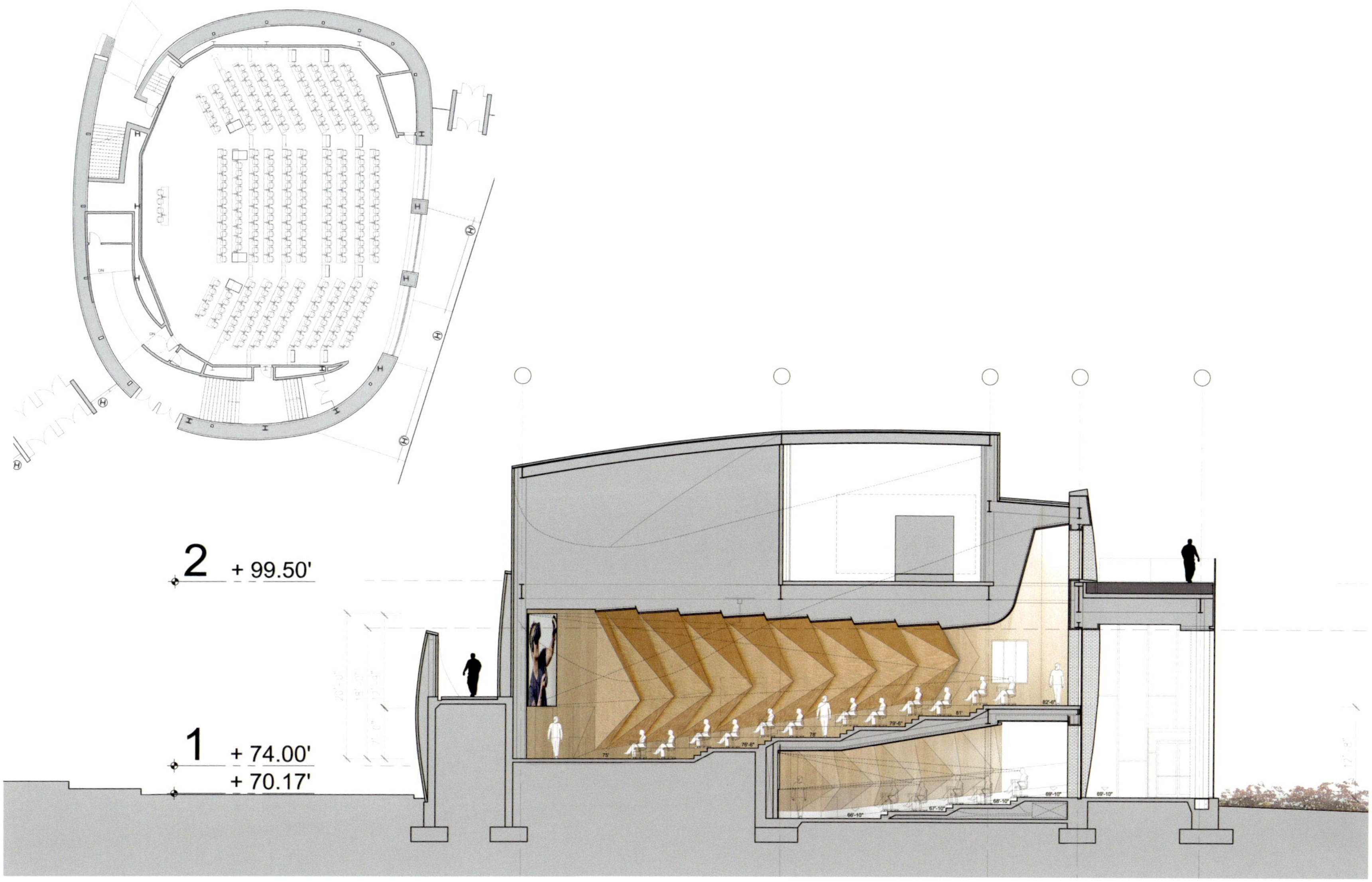

Plan and section

1

Institute for Data Science

Kennedy & Violich Architecture

Location University of Rochester, New York
Function Higher education and scientific research
Client University of Rochester
Completion date 2019
Architect Kennedy & Violich Architecture
Landscape CRJA-IBI Group
Structural Engineer Buro Happold Limited
MEP & Envelope IBC Engineering
Civil Engineer T.Y. Lin International
Contractor Welliver
Photography John Horner, courtesy of KVA

The new Institute for Data Science at the University of Rochester is the signature focus of the University's Strategic Plan. The plan is intended to foreground information technology in a new educational model for cross-disciplinary research and innovation. The 61,000 square foot building program includes smart classrooms and faculty offices, robotics, science, computation and environmental labs, a 160-seat auditorium, and the Goergen Institute, a new public resource for information technology.

Digital brick To meet the research mission of the Institute for Data Science, the design takes a somewhat different conceptual and constructive approach to brick. The project transforms the University's standard red brick in a design that treats each brick as a digital 'pixel' of information capable of expressing its status in relation to the building's overall geometry in the campus context. The taut brick skin is subdivided into running and Flemish bond patterns with projecting header bricks in the Flemish bond zones, designed using generative algorithms visualized in digital 3D modeling software.

2

A design that treats each brick as a digital 'pixel' of information.

1 Campus plan with river view corridors.
2 Textured brick, tied to building geometry data cloud.
3 South façade from Engineering Quad.
4 Campus entry stair and data platform.
5 River entry collaboration breakout space.

3

Shadows from the projecting bricks activate the façades, creating dynamic, textural surfaces.

4

This creates 'data streams' of textured brickwork which provide a reading of the massing's inflections to the river and the campus. Shadows from the projecting bricks activate the façades, creating dynamic, textural surfaces that communicate natural environmental data across the days and seasons—information made legible at multiple scales across the brick surfaces of the building. Narrow and wide punched windows are inspired by the binary 0/1 that is fundamental to encoding data.

Social collaboration spaces In the interior, social collaboration spaces are integrated into a generous public circulation sequence, which follows the building perimeter and engages each floor level with double-height collaboration zones with views to the Genesee River and the Science & Engineering Quad. The sequence of collaboration space culminates at a double-height, shared lab space with skylights that overlook the Quad and is linked by a new bridge to neighboring Hopeman Hall. Lab spaces are placed adjacent to collaboration spaces, using a modular bay system which supports flexibility and interdisciplinary research projects. All interior lab corridors terminate with window overlooks which admit natural light deep into the building interior. The public circulation sequence is expressed on the exterior façades, with large vertical glazed bays that 'breakthrough' the brick envelope to reflect double-height collaboration zones, and shared overlooks that admit natural light deep into the building interior. A double-height curtain wall on the first and second level interlocks with the figure of hung brick on the upper floors, expressing the Institute's mission to create an open and collaborative research culture that engages the green space of the renewed Science & Engineering Quad.

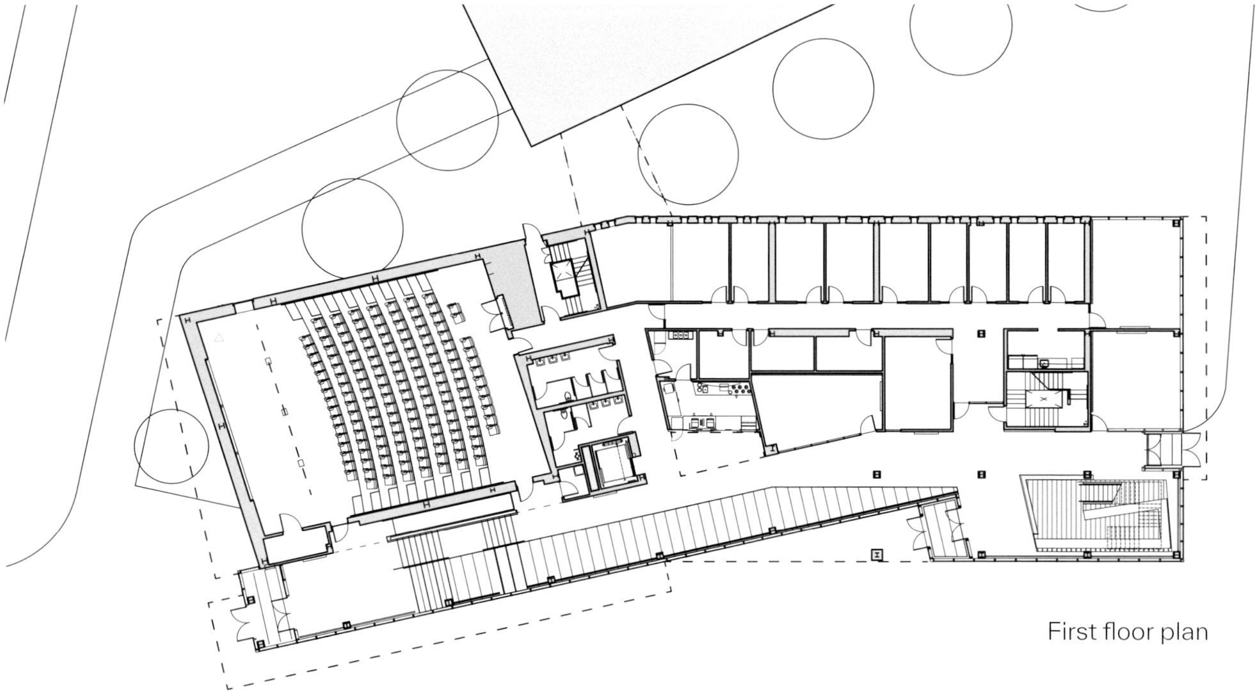

First floor plan

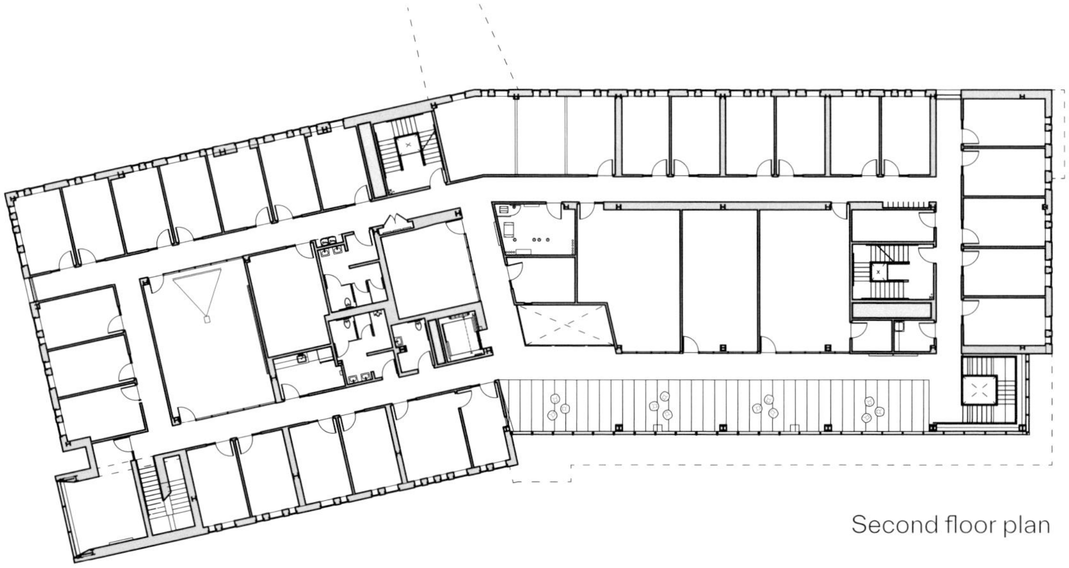

Second floor plan

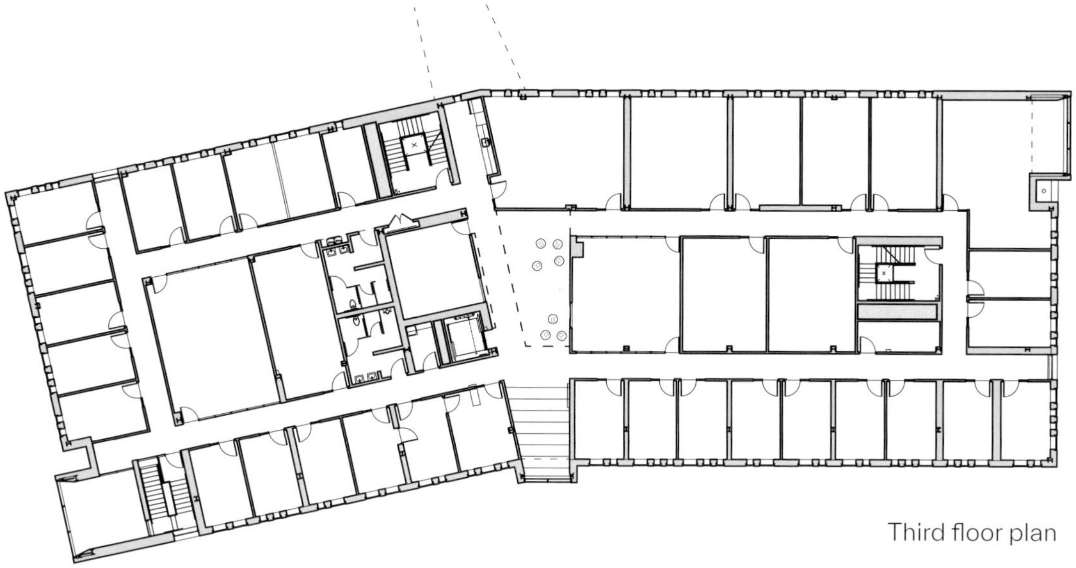

Third floor plan

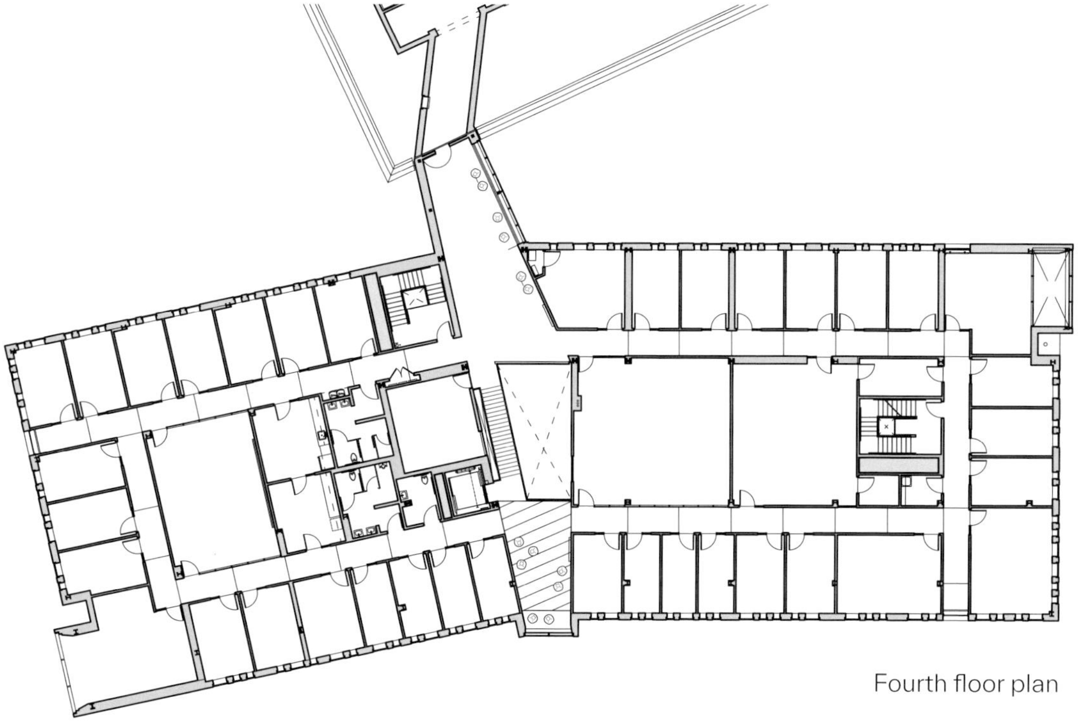

Fourth floor plan

Tozzer Anthropology Building

Kennedy & Violich Architecture

Location Cambridge, Massachusetts
Function Anthropology building
Client Harvard University
Completion date 2014
Architect Kennedy & Violich Architecture
Sustainability LEED Gold certification
Landscape Richard Burck Associates, Inc.
Structural Engineer Lemessurier Consultants
MEP & Envelope Buro Happold Limited
Lighting Designer Tillotson Design Associates
Photography John Horner, courtesy of KVA

Modernist era university buildings are aging across America. This design transforms the public presence and programs of Harvard University's 1971 Tozzer Library while re-using the existing building's foundation, existing structure, and campus infrastructure connections. The design houses the new Anthropology Library, faculty offices, collegial spaces, and classrooms organized around a 3-story day lighting atrium. The new massing adds usable space with a copper roof volume, which captures daylight and strengthens a reading of the building as a pavilion in the Peabody Courtyard.

The original 1971 brick envelope could not be reused because it did not meet current energy performance criteria. Using contemporary brick veneer construction, the design invents a digital brick corbel that operates spatially at the scale of the building to create a distinctive 'thick', sculpted public entry. The design creates a set of bespoke algorithms that link the form of each standard modular brick to the overall geometry of the massing. This enables the new brick façade form and volume to be a direct expression of the

1

2

building's geometry and demonstrates how a contemporary hung brick wall could become authentic to its building form. A tilted metal stud back-up is cantilevered at the second floor allowing the 30' brick entry to be constructed without interruption of control joints or relieving angles, in one carefully designed loose lintel set above the hung brick ceiling.

Design collaboration The digital 'corbel' was 3D modelled and tested in a unique collaboration process with the design team and local union masons. The brick façade design was mocked up through physical dry stack prototypes, integrating digital and physical forms of knowledge. The main building volume is wrapped by 'thin' taut brick bands which express the existing slab, from which the

3

The digital 'corbel' was 3D modelled and tested in a unique collaboration process with the design team and local union masons.

new brick envelop is relieved. Vertical expansion joints are staggered between floor levels while horizontal expansion joints occur behind recessed soldier courses. The structurally glazed projecting windows slide past the 'thin' brick envelop and align to provide generous views from Library to Courtyard. With a play of 'thick' and 'thin', this design collaboration reflects contemporary construction, existing reuse circumstances and the historic brick content of the adjacent Harvard Peabody Museum and campus context.

Brick carbon bank The design of the Tozzer Anthropology building takes a new approach to adaptive reuse and sustainability by treating the building as a carbon bank that expands time scales in architecture.

On the interior, the use of sustainable materials meets all consumer content recycling requirements including FSD certified woods and re-cycled (100 Year) copper roof. Day lighting and efficient active chilled beams and a heat recovery system reduce cooling loads and conserve operational energy.

The design preserves and amortizes the embodied energy of the building's existing structure and infrastructure and enables the addition of an enduring new brick envelope, which in time, could also be replaced. The design banks embodied energy and enables the periodic exchange of new masonry building envelopes, responding to contemporary needs for both permanence and change. This innovative project of adaptive reuse opens new possibilities for local material sourcing, supports local economies and collaborations with skilled constructors and masons and provides significant carbon savings in embodied and operational energy expenditures over time.

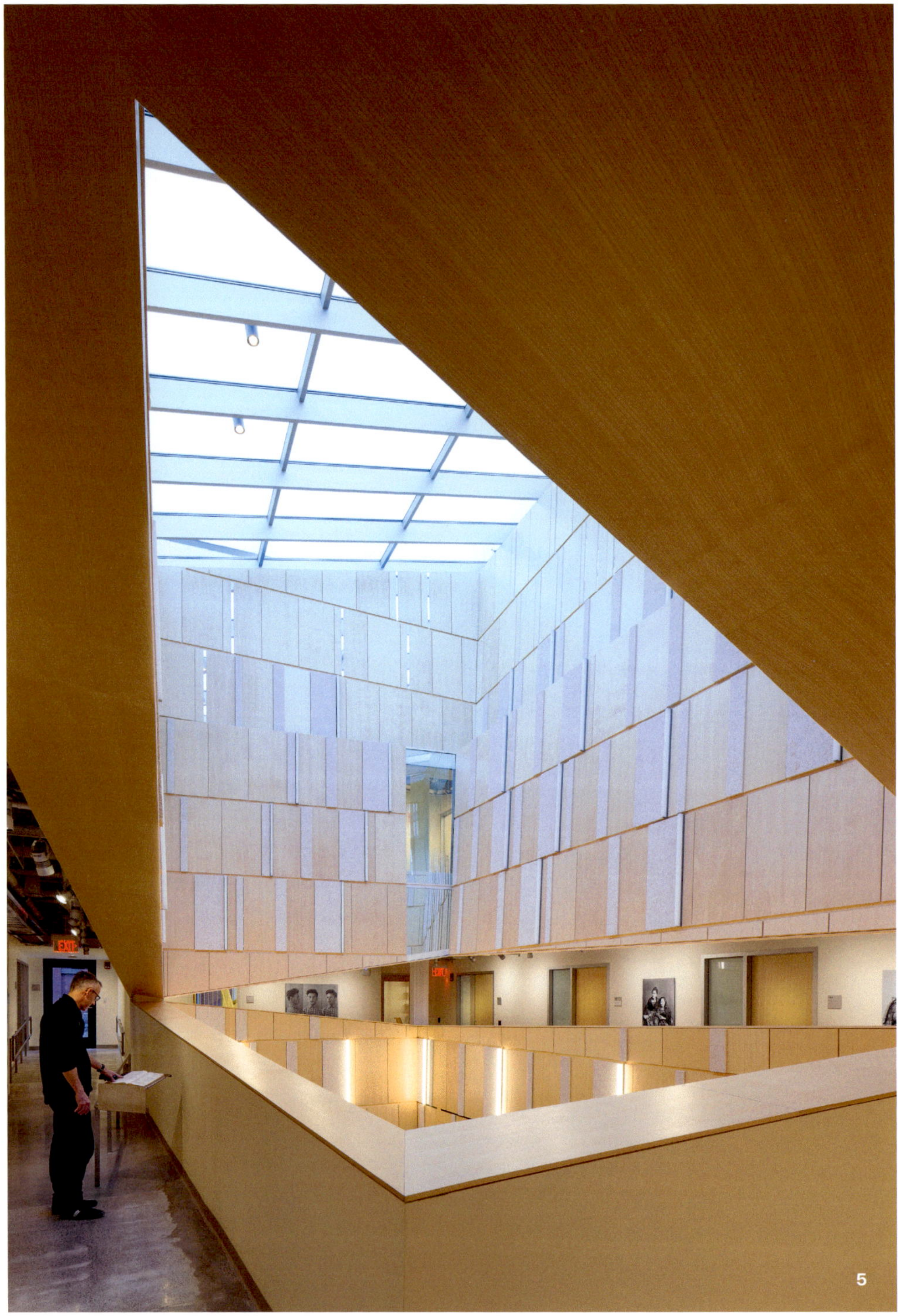

5

1 View of the main entrance from Divinity Ave.

2 The second-floor collegial space opens to the light-well above, serving as a central 'living room' for the department.

3 Brick Entry Pavilion reads as a thick masonry structure.

4 View looking north-east from Divinity Ave.

5 View of the light-well from the third floor.

6 The library windows face the Peabody Courtyard, which features historic stone artifacts.

The Tozzer Anthropology building takes a new approach to adaptive reuse and sustainability.

6

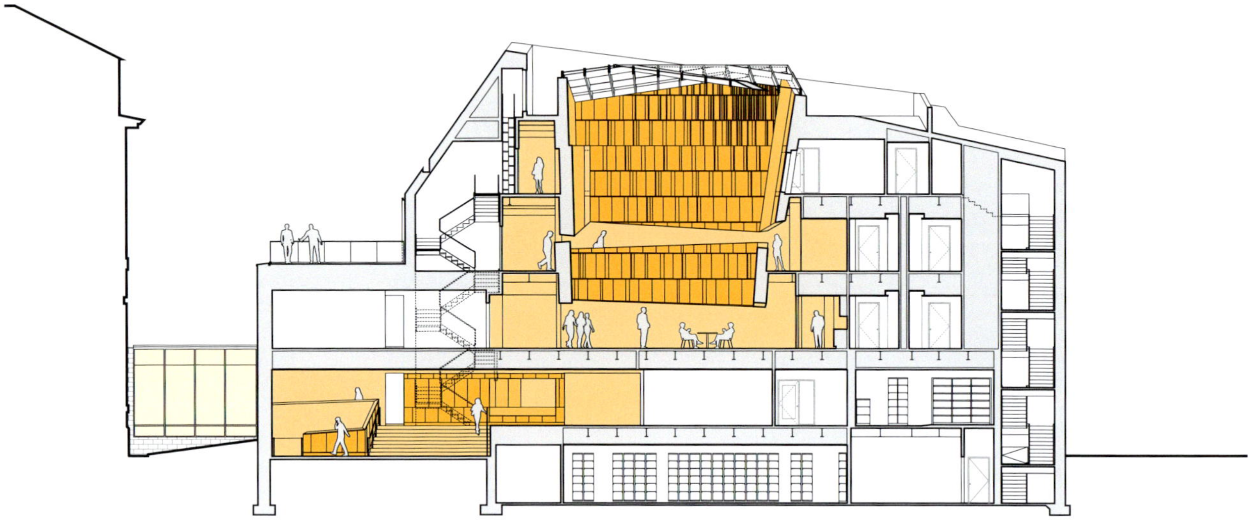

North/south section

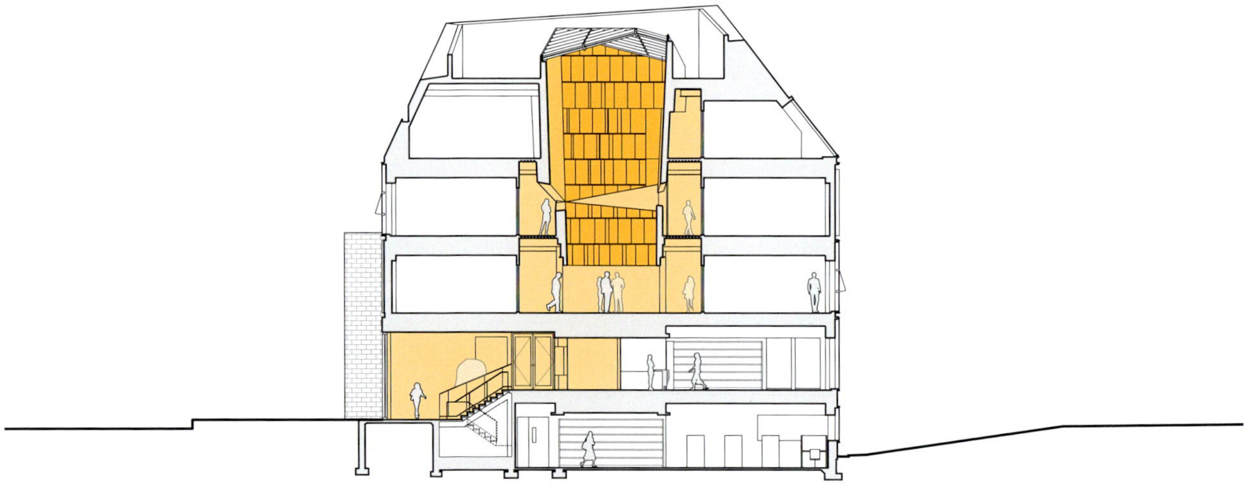

East/west section

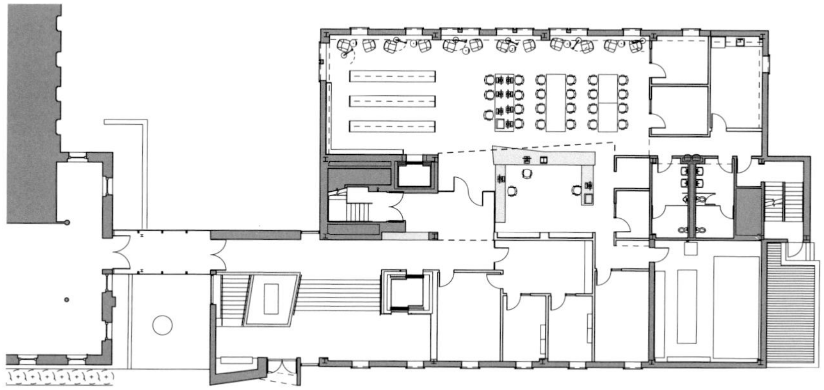

First floor plan

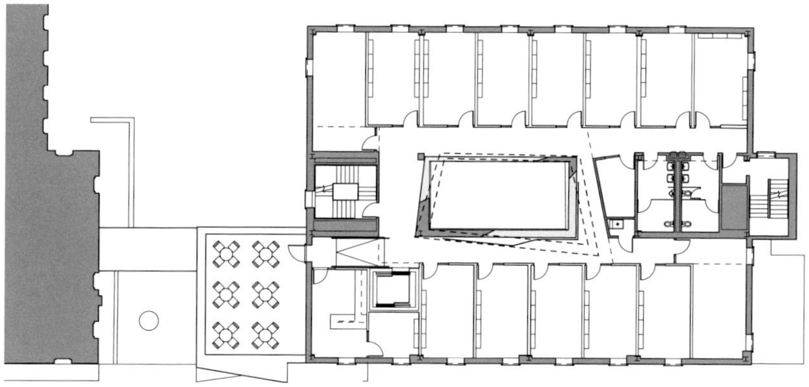

Third floor plan

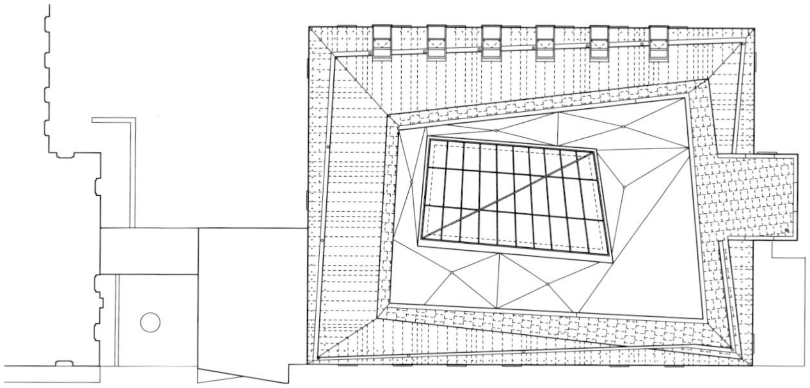

Roof plan

Wilbur O. & Ann Powers College of Business

LMN Architects

Location Clemson, South Carolina
Function College of Business
Client Clemson University
Completion date 2020
Architect LMN Architects
Associate Architect LS3P Associates Ltd
Landscape Ayers Saint Gross
Engineer Land Planning Associates, Inc.
Contractor DPR Construction
Photography Mark Herboth Photography

1

Respecting the historical significance of the location was a fundamental requirement.

The vision for a business school goes beyond traditional methods, aiming to prepare students for effective collaboration on personal, organizational, and global levels. Clemson University recognized the need for its new College of Business building to embody a fresh way of learning, with innovative spaces tailored for the evolving future. Positioned by the historic Bowman Field, a longstanding campus feature where the university's inaugural sports event took place in 1896, the building holds a significant place in the campus landscape.

Preserving heritage and connectivity Respecting the historical significance of the location was a fundamental requirement. As the first academic structure in this section of the campus in more than a century, the building had a dual role: to establish a new entry point to the campus and to enhance the connection between the university and the surrounding city. This rare opportunity not only provided a modern setting for business programs but also accommodated diverse learning approaches while acknowledging the institution's heritage. The Dean's primary objective was to create an environment so appealing that students would be reluctant to leave.

The prominent feature of the site, a 40-foot hill between Bowman Field and the student residences above, presented both challenges and possibilities. Leveraging the striking topography, the design team conceived the hill ascent as a central element reminiscent of the Spanish

A grand interior staircase mirrors the hill's slope, inviting students to enter the building at different levels as they traverse the terrain.

Steps. The structure took on a shape that adapted to the hill's incline, connecting to a linear section. This design framed outdoor terraced stairs and plazas. The building's materials were chosen to echo the hill's character, featuring sturdy concrete elements and natural-toned tiles at the base, transitioning to lighter, airier materials as the structure ascends. Inside the atrium, a lattice of white vertical slats draws attention upward. A grand interior staircase mirrors the hill's slope, inviting students to enter the building at different levels as they traverse the terrain.

Balancing design and purpose The building's form and detailing harmonize with the arrangement of the student residences above. Students move through a planned sequence that guides them from the lower plaza to terraced communal spaces and finally to a promenade at the summit, interconnecting living, academic, and social areas. The expansive atrium is carefully designed to foster connections in various ways. It includes gathering spots, adaptable furnishings, and convenient spots for brief tasks.

The project significantly boosted the college's reputation, highlighted by the MBA program's recognition as one of the top 30 by the Princeton Review. Moreover, a recent record-breaking $60 million contribution, the largest in the university's history, was made to support scholarships and programs. On a larger scale, the once-remote site is now a vibrant activity hub for the campus. By consolidating programs from five separate buildings into one, while doubling the college's previous footprint, the new facility enhances community interaction and streamlines operations. Dean Wendy York emphasizes that the building, aligned with the academic vision, serves as a launching pad for students to become catalysts of positive change in South Carolina's business landscape, across the nation, and even beyond.

5

1 Sited at the highest point in the area, the Wilbur O. and Ann Powers College of Business creates a new gateway for Clemson University.

2 The design team conducted research on the history, architectural context, proportions, fenestration of nearby buildings, and other elements to inform a thoughtful design intervention on the site.

3 Capitalizing on a 40' hill between Bowman Field and the student residences above, the team envisioned the hill climb as a defining feature akin to the Spanish Steps in Rome.

4 The atrium is designed for connection with gathering areas, adaptable furnishings, and touchdown spaces to accommodate a variety of tasks.

5 The interior palette is defined by natural materials that echo the solidity of the mountains and celebrate the horizontal strata of the hillside.

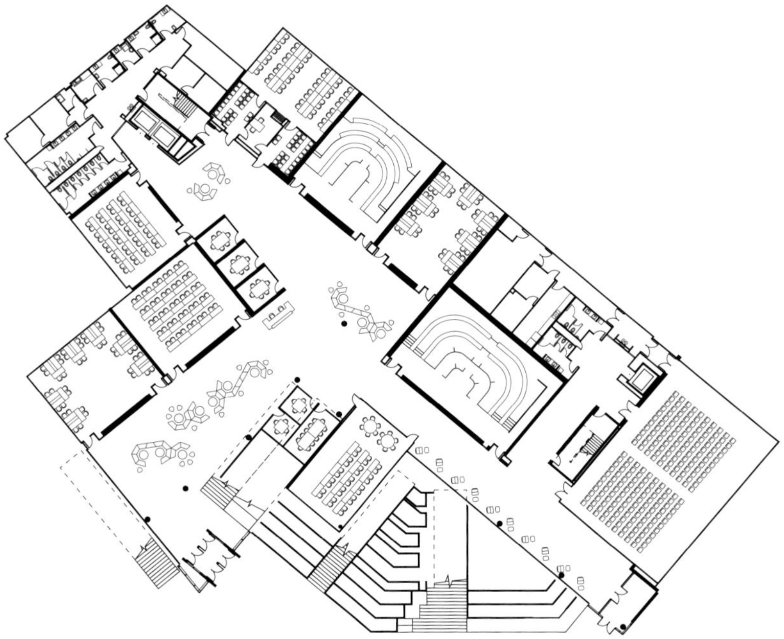

Ground level plan

Level one plan

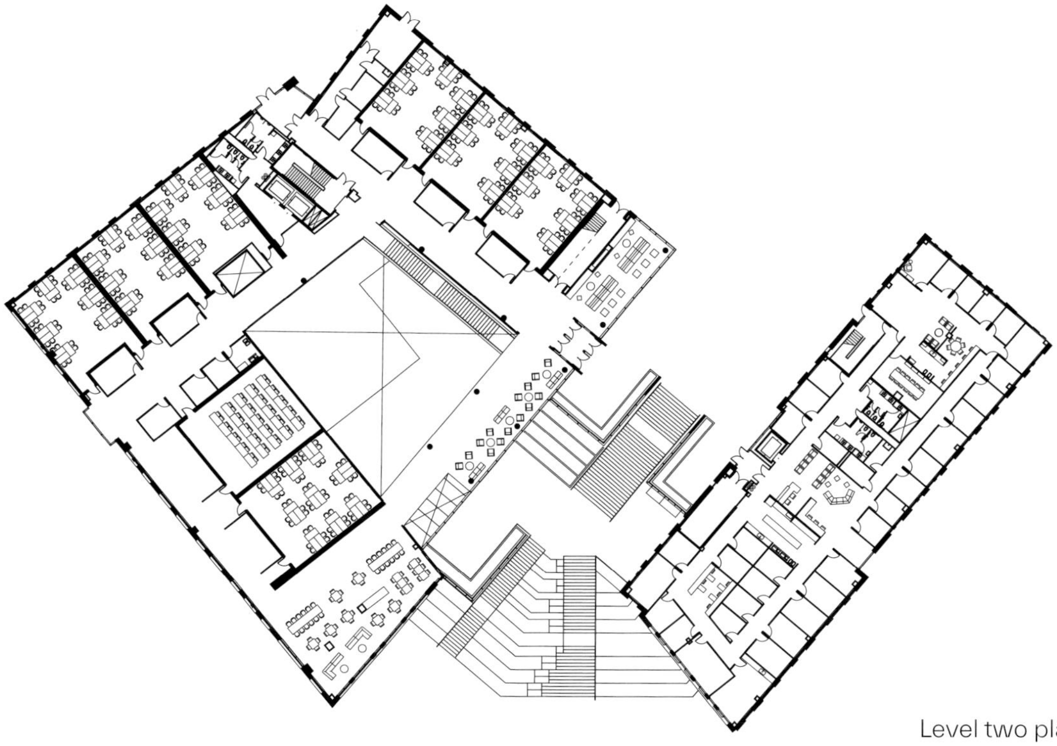

Level two plan

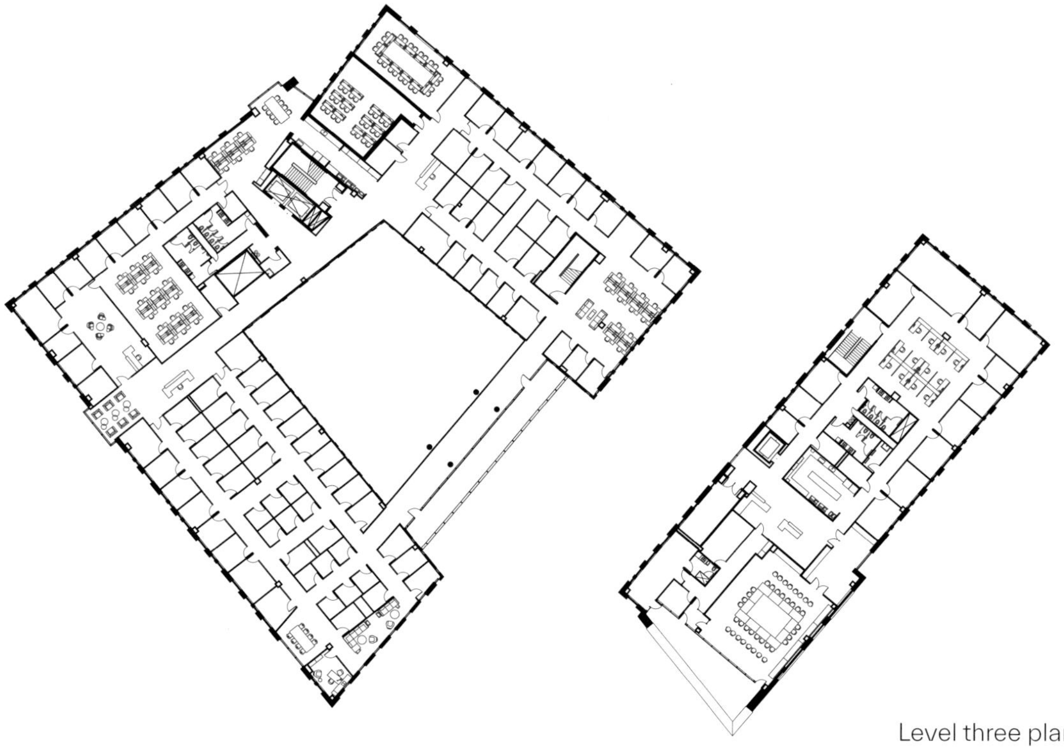

Level three plan

The Ohio Union
Moody Nolan

Location Columbus, Ohio
Function Student Union
Client The Ohio State University
Completion date 2010
Architect Moody Nolan
Size 31000 ft^2
Sustainability LEED Silver Certified
General Contractor Smoot Construction
Landscape The EDGE Group
Photography Brad Feinknopf/OTTO

1

Moody Nolan played a significant role in The Ohio State University's project to build a new student union right from the beginning. They collaborated with the university to conduct an initial analysis and feasibility study. As the project progressed and shifted from an idea to an actual plan, Moody Nolan became the Executive Architect in partnership with the design architect, Michael Dennis. They took over full design and production responsibilities after the Design Development phase.

The University aimed to construct a distinctive Student Union building that honored its historical masonry surroundings while also presenting a modern design for future generations of students. The Union holds a prominent position on the campus, acting as a gateway from the nearby city street and serving as a central point for student activities. The building was carefully designed so that its functional needs, like dining and event services, were discreetly integrated within a parking structure, maintaining the brick and masonry aesthetic.

Inclusion and involvement The planning, programming, and design process deliberately involved a wide range of participants. This included representatives from the University, students, alumni, and the local community who were brought together through focus-group sessions. Additionally, a university-wide questionnaire was distributed to encourage even more student engagement. The insights from these sessions and surveys played a crucial role in shaping the design of the new Ohio Union.

2

The building features three major façades, each serving a distinct purpose: the city-facing side boasts a welcoming entry arch and a large ballroom window overlooking the street, the south side has an event courtyard and an entry arcade, and the campus side features a recessed plaza that serves as a backdrop to a campus green space. The exterior, mainly made of brick with cast stone accents, blends harmoniously with the campus environment. Multiple entrances along the perimeter lead to a central atrium, providing simple navigation and a hub for student activities. The atrium, adorned with the University Alma mater inscription, spans three levels and houses balconies, stairs, and overlooks. Interior brickwork highlights student organization commons, lounge areas, and collaborative spaces, including fireplaces. The two ballrooms, theater, and atrium accommodate significant campus events.

The building successfully combines contemporary and historically inspired designs, creating an iconic structure that feels integral to the campus.

3

Materiality Brick and masonry materials were chosen to allow for diverse architectural approaches across the different façades, accommodating various entryways and window configurations. The building successfully combines contemporary and historically inspired designs, creating an iconic structure that feels integral to the campus. The use of brick also provides a backdrop for stone art reliefs depicting University life. Some of these reliefs were reclaimed from the previous 1950s Union on the same site, while others were modern additions, promoting diversity and inclusion on campus. The building's structural framework is made of steel with steel stud backup, supporting its LEED Silver certification. The façade's porosity allows natural light to permeate the interior, enhancing the overall user experience in rooms and circulation areas.

The new Ohio Union stands as one of the largest student unions in the country. It offers a wide range of amenities, including a food court and standalone food/beverage options, performance spaces of various sizes, a versatile ballroom, advanced conference rooms, and well-equipped student organization offices. These facilities are complemented by numerous lounge and gathering spaces, essential for a vibrant and multifunctional student union.

1 Anchored on the eastern edge of campus, the Ohio Union establishes a contemporary identity while remaining contextually attuned to the adjacent South Oval lawn and neighboring brick buildings.

2 Lounge spaces are interspersed throughout the Union, offering students comfortable and quiet study spaces within the bustling campus hub.

3 Spans of glazing and vertical mullions delineate the Union's expansive brick façades in a rhythmic arrangement.

4 A fifties-style diner activates the Union from early morning to late at night. Pops of scarlet and gray are woven into material finishes, and furnishing complements the vintage university emblem.

5 Clerestory windows illuminate the Great Hall with ample daylight, benefiting surrounding program spaces.

Ohio UNION
WELCOMING HOME

West elevation

East elevation

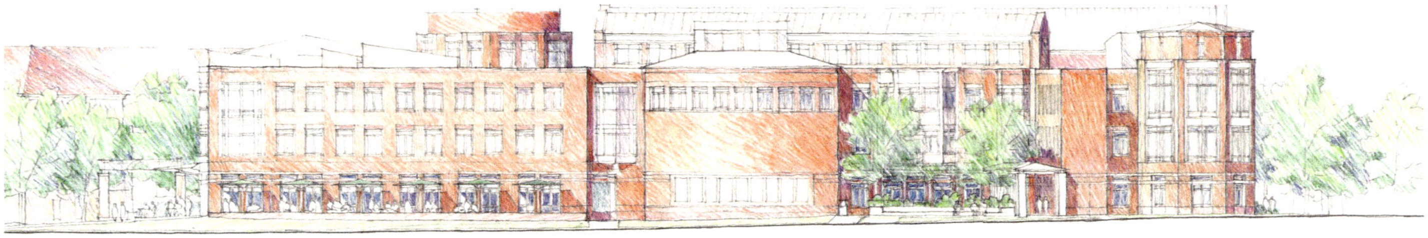

South elevation

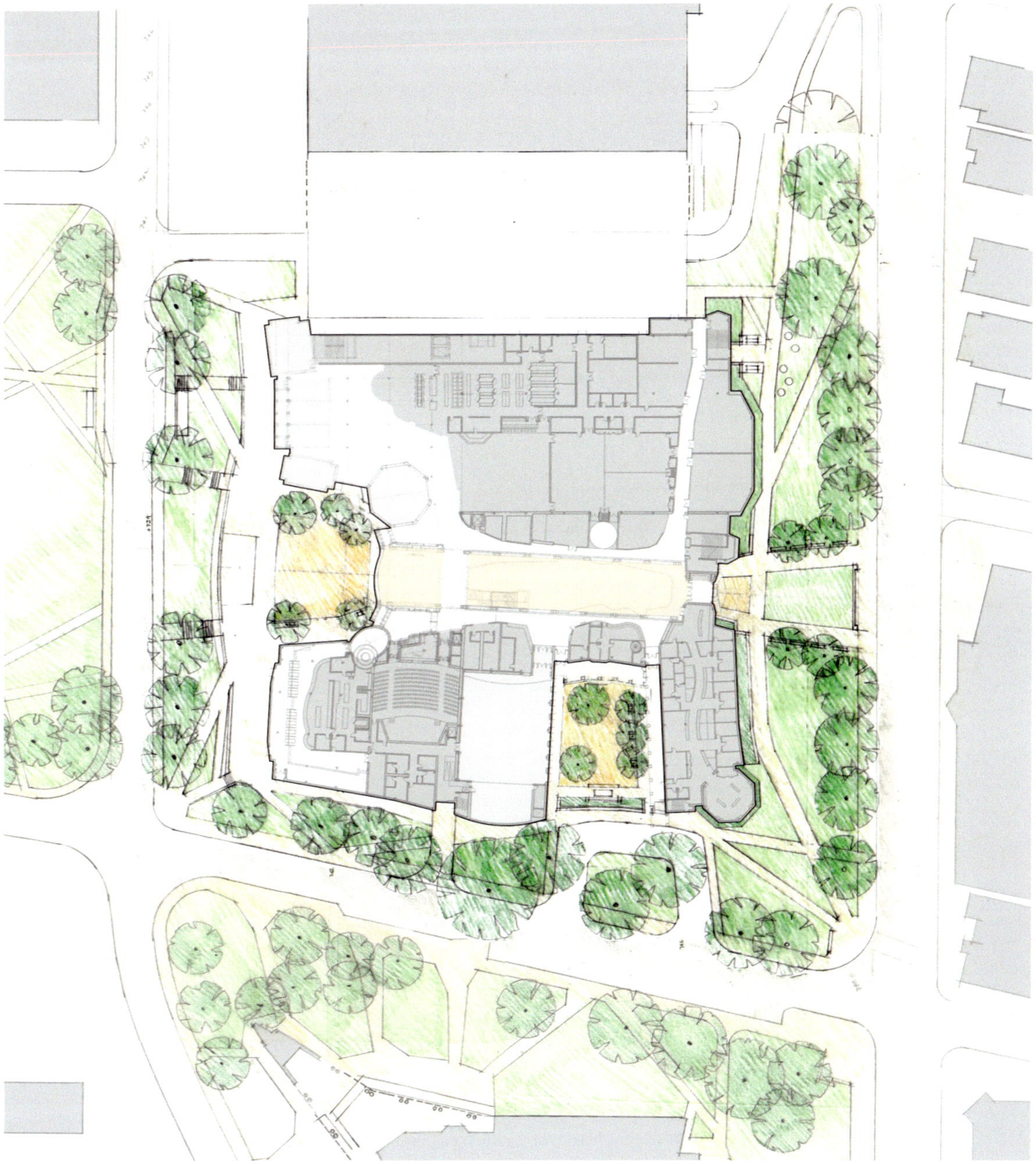

Site plan

1

Wilf Hall

MA | Morris Adjmi Architects

Location New York, New York
Function Offices and Community Theater
Client New York University School of Law
Completion date 2010
Architect MA | Morris Adjmi Architects
Size 50,000 ft²
Sustainability LEED Platinum
Structural Engineer Robert Silman Associates
Mechanical Engineer FMC Engineering
Environmental/civil engineer Langan
General Contractor Skanska
Landscape Architect Mathews Nielsen
Theater consultant Robert Lorelli Associates
Lighting designer RS Lighting Design
Façade/waterproofing consultant Frank Seta & Associates
Photography Jimi Billingsley & Matthew Williams

Wilf Hall's most significant sustainable feature is its façade, which is composed of bricks made using recycled industrial materials.

2

Located in the heart of Greenwich Village, Wilf Hall is a new six-story building that provides New York University School of Law with contemporary offices and workspace for more than a dozen academic programs while respecting the character and cultural legacy of the beloved historic neighborhood.

Recalling the original 1830s brownstones that once occupied the site, Wilf Hall's brick façade is divided into four nearly identical rowhouse-sized bays by three narrow 'slot' windows. A fine, imperturbably smooth red brickwork extends over the entire façade. That flatness is relieved to subtlest effect by the geometric regularity of the building's sash windows and a demure and continuous metal cornice on the top. Breaking up the long wall surfaces adds visual richness, and it also signals the structure's historical presence. Narrow, aluminum-framed, and recessed glazed apertures extending from ground to cornice neatly accomplish both objectives.

The building's southern bay incorporates the renowned Provincetown Playhouse—one of the most important sites in the history of 20th century American theater, which served as a launching pad for works of Eugene O'Neill, Edna St. Vincent Millay, Edward Albee, John Guare, Sam Shepard, Charles Busch, and David Mamet, among others. The theater was respectfully restored and upgraded to ensure its legacy continues, and its original brick detailing continues into the new structure to help mitigate the building's scale and further anchor it to the neighborhood.

Set back to maintain the existing cornice line, Wilf Hall's penthouse is covered by a green roof. However, the building's most significant sustainable feature is its façade, which is composed of bricks made using recycled industrial materials, such as batteries, previously destined for a landfill. The project was the first education building in New York to earn LEED Platinum certification.

3

The building's southern bay incorporates the renowned Provincetown Playhouse—one of the most important sites in the history of 20th century American theater.

1 To fit the scale of Greenwich Village, Wilf Hall was designed to resemble four brick townhouses.

2 Keeping the existing context in mind, the building's penthouse level is set back to help preserve the smaller scale of the neighborhood and maintain an existing cornice line.

3 The penthouse set back creates private outdoor amenity space for the building's occupants to enjoy.

4 A large swath of green roof helped the project achieve LEED Platinum certification.

5 The NYU School of Law academic building incorporates a preserved the historic Provincetown Playhouse theater into its façade.

5
The PROVINCETOWN
133
131

Elevation

Level one plan

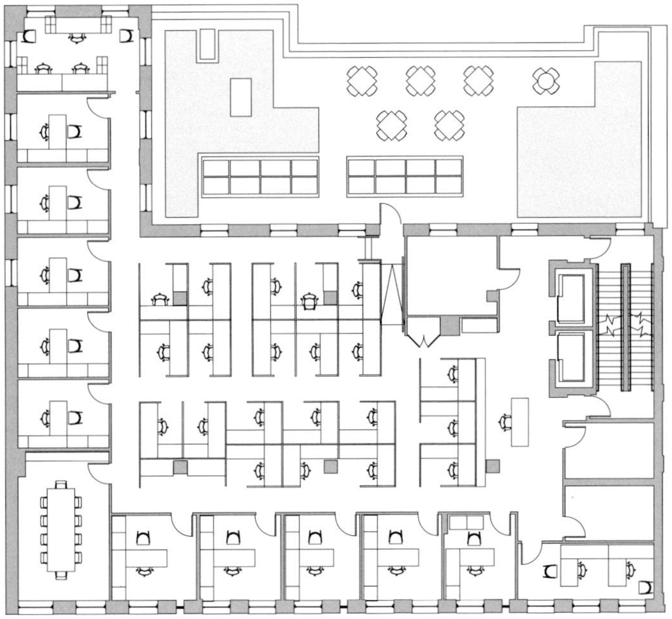

Level two plan

1

RISD North Hall

NADAAA

Location Providence, Rhode Island

Function Higher education dormitory and art studio

Client Rhode Island School of Design

Completion date 2019

Architect NADAAA

Landscape Landworks Studio

Structural Engineer Odeh Engineers

Civil Engineer Vanasse Hangen Brustlin Inc. (VHB)

Mechanical Environmental Systems Inc.

Contractor Shawmut Design and Construction

Photography John Horner

This project involved creating a new residence hall at the Rhode Island School of Design, and was the first step in a series of renovations to improve RISD's Freshman Quad. The goal was to make better use of public outdoor spaces and tie the surrounding buildings together. The client's mission was to maximize the number of beds, which led to a larger building than originally planned for the site. Next to the List Building designed by Philip Johnson, the approach for this new building was to carve out as much space from its mass as possible. This helps connect its different parts—such as cornice lines, terraces, and materials—to the nearby surroundings on Angel, Prospect, and Waterman streets. The building also includes small studio spaces within its break-out areas, creating a friendly and social atmosphere for art and design students.

North Hall's materials were chosen to match the nearby surroundings. On the north and south sides, a stacked bond brick façade is used to visually connect the building to the site. The indented bricks add depth and solidity. On the east and west sides, a lightweight cement-fiber façade

The goal was to make better use of public outdoor spaces and bring together the surrounding buildings.

highlights the local vernacular shingle design seen on the RISD and Brown campuses. The east and west sides are accentuated by awning windows that not only enhance the shingle appearance but also function effectively, highlighting a sense of lightness. Mass timber slabs made from cross-laminated timber (CLT) replace concrete, lowering CO_2 emissions by up to 20%.

A light gray cool roof membrane, Sika Sarnafil, EnergySmart, meets environmental criteria by helping reduce heat island effects. Custom furniture and counters are crafted from field-cut CLT slab pieces. Exposed ceilings in dorm rooms, common areas, and corridors eliminate the need for suspended ceilings and cut down on energy use. The custom Rhye furniture collection used in the building is made from just two sustainable materials: PEFC-certified solid European beech and fast-growing, carbon-neutral bamboo plywood.

Brick as structural element A key urban feature in Providence is the RISD Beach with its distinctive brick façade created by Pietro Belluschi. This brickwork ties together Benefit Street, Waterman Street, and Angel Street with downtown. This brickwork serves as the foundational element of the site, functioning as retaining walls, pilasters, and façades on the north and south sides of North Hall. It visually connects the extension to Prospect Street and the New Dormitory up the hill. The brick design is anchored on the south side by Nickerson Court's hard surface and on the north side by the open Nickerson Green, separated by the new RISD breezeway.

4

Advancing sustainability An important aspect of this project is incorporating building technologies that optimize material usage. By using a combination of steel and cross-laminated timber, the project achieved 9-foot high raw wood ceilings without extra finishes. This design allows mechanical systems to be placed in the corridors, minimizing their impact on bedrooms. The bedroom itself helps circulate fresh air through the mechanical system, working alongside the variable refrigerant flow unit that controls the air temperature. By combining thoughtful design, sustainable materials, and innovative systems, the new residence hall at RISD showcases a harmonious blend of practicality, aesthetics, and environmental responsibility.

Exposed ceilings in dorm rooms, common areas, and corridors eliminate the need for suspended ceilings and cut down on energy use.

5

1 Bricked and shingled façades.
2 North façade and Nickerson Green.
3 Brick detail on south façade.
4 East façade.
5 Shared studio.

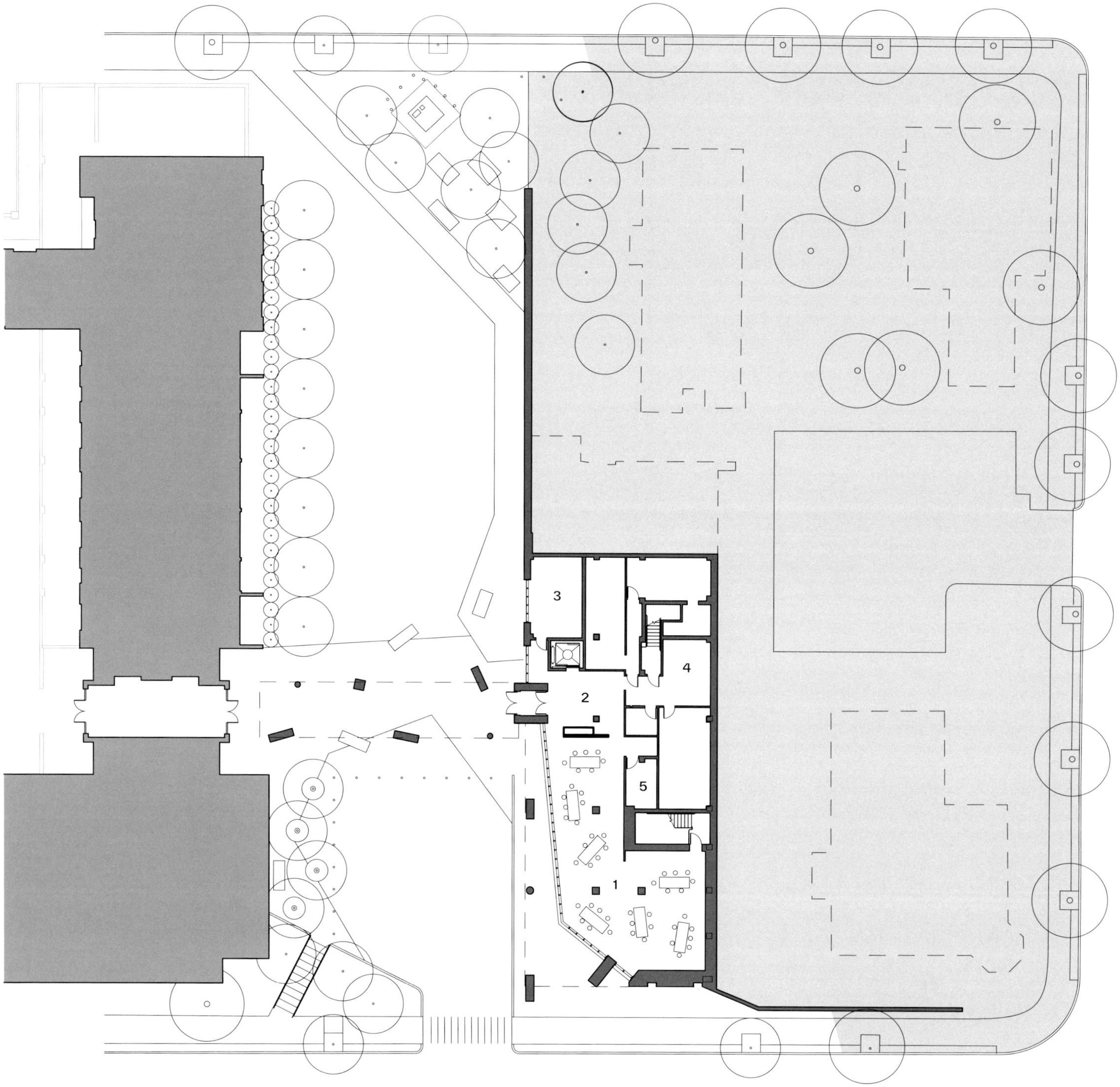

1 Studio
2 Lobby
3 Laundry
4 Bike storage
5 Spraybooth

Level one

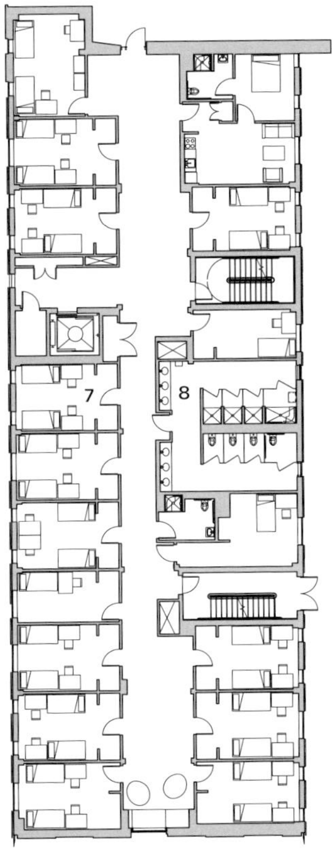

Level two (typical)

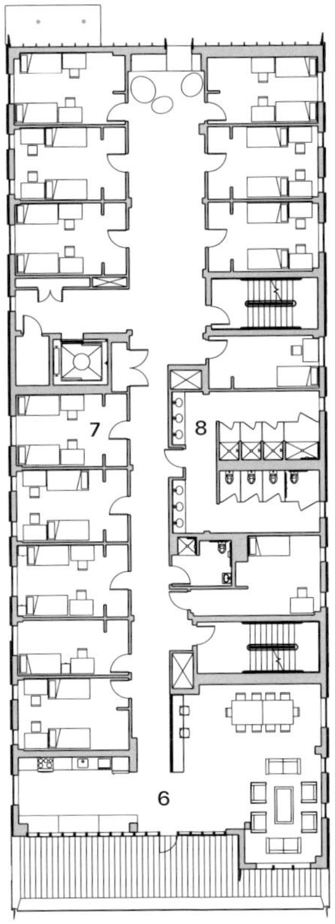

6 Lounge
7 Dorm room
8 Bathroom

Level five

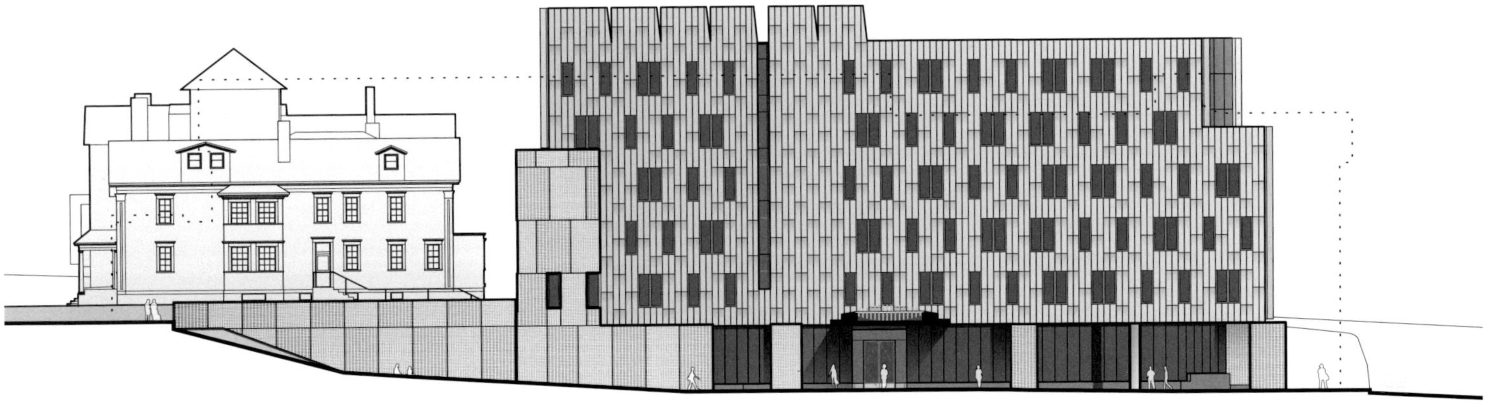

West elevation

NYS Equal Rights Heritage Center

nARCHITECTS

Location Auburn, New York
Function Heritage Center
Client City of Auburn
Completion date 2018
Architect nARCHITECTS
Sustainability Geothermal Heating and Cooling
Landscape EDR and nARCHITECTS
Consulting/Civil Engineer EDR
Structural Engineer Silman
MEP Engineer OLA Consulting Engineers
Contractor Fahs Construction
Lighting Lumen Architecture
Photography James Ewing/OTTO and nARCHITECTS
Drawings nARCHITECTS

The City of Auburn and the New York State Office of Parks, Recreation and Historic Preservation commissioned nARCHITECTS to create the New York State Equal Rights Heritage Center and a public outdoor space in historic Auburn, NY. The 7,500-square-foot building serves as a permanent exhibition space that celebrates the progressive history of promoting social and equal rights within the City and the State. The exhibition content is centered around various civil rights movements, including abolition, women's suffrage, and LGBTQ rights. The objective is to introduce visitors to the State's diverse array of attractions associated with the concept of equal rights.

The project aimed to bolster tourism in the region by establishing a welcoming center for the numerous historical sites in the vicinity. The building is situated near notable landmarks such as the Seward House Museum, Memorial City Hall, the historic YMCA, and Westminster Church. The interior layout of the exhibition spaces is designed to guide visitors along the building's perimeter, where large windows offer views of these significant local landmarks.

Striving to blend into Auburn's historical district and downtown area while forging meaningful links between the past and present, the Center's design exhibits a thoughtful contemporary approach. Angled brick volumes are skillfully

1

Angled brick volumes thoughtfully rotate and intersect to frame views.

arranged to rotate and intersect, simultaneously connecting with the historical surroundings and capturing scenic views at key junctures. The permanent exhibition—co-designed by nARCHITECTS and MTWTF—incorporates interactive elements that vividly portray the struggle for equal rights through music, speeches, games, portraits, posters, animated maps, and video timelines. The 'Social Justice Table' is a standout element that brings people together around a circular video display highlighting legislative milestones. Conceptually, the center is envisioned as Auburn's communal space—a versatile indoor and outdoor area accommodating programs that adapt to the evolving needs of the community. Within the building's four volumes, amenities such as a Taste NY store, public restrooms, and offices for the Auburn Historic & Cultural Sites Commission and the local Business Improvement District are included. Visitors experience the permanent exhibition while remaining connected to the historic essence of Auburn that envelops them.

Reflecting its historical context, the single-story building's design places significant emphasis on proportionality, featuring ample glazing that conforms to the building's structure. Angled brick volumes thoughtfully rotate and intersect to frame views that celebrate both the surrounding landscape and noteworthy historical landmarks. A deliberately restrained selection of materials includes a rose-colored brick, chosen to harmonize with the historic houses on South Street. Simple yet impactful details, such as angled brick coursing and corbels at window jambs, contribute depth and texture to the facade.

The building's luminous interior showcases exposed, resilient structural materials such as board-formed concrete walls and a timber-framed roof. Meticulous attention was paid

3

to the proportions of the steel windows, the details of the brick facade, and the copper-colored anodized aluminum roof, all with an eye toward preserving a connection to the diverse historical context. Geothermal heating was employed to ensure energy efficiency while maintaining unobstructed roof lines, devoid of visible mechanical equipment. The most notable challenge was the tight schedule, as the Center was conceived, designed, and constructed in just 18 months, with a firm opening date set from the beginning. The design team made strategic choices in terms of structural systems, materials, and fabrication techniques to meet this demanding timeline. In addition to adhering to an ambitious schedule, efforts were made to maximize value for the community. Remarkably, the project was completed $1 million under budget, with the excess funds being allocated to the City's annual operating budget.

1 NYS Equal Rights Heritage Center with a statue of Harriet Tubman. Image: James Ewing.

2 Looking into the Center at the Key Figures in New York State Equal Rights History Exhibit. Image: James Ewing.

3 The Center's brick façade. Image: nARCHITECTS

4 Speech Island, an exhibit where visitors can rest and listen to historical speeches about equal rights. Image: James Ewing.

5 The Equal Right Heritage Center frames views out as well as focusing attention within. Image: James Ewing.

5

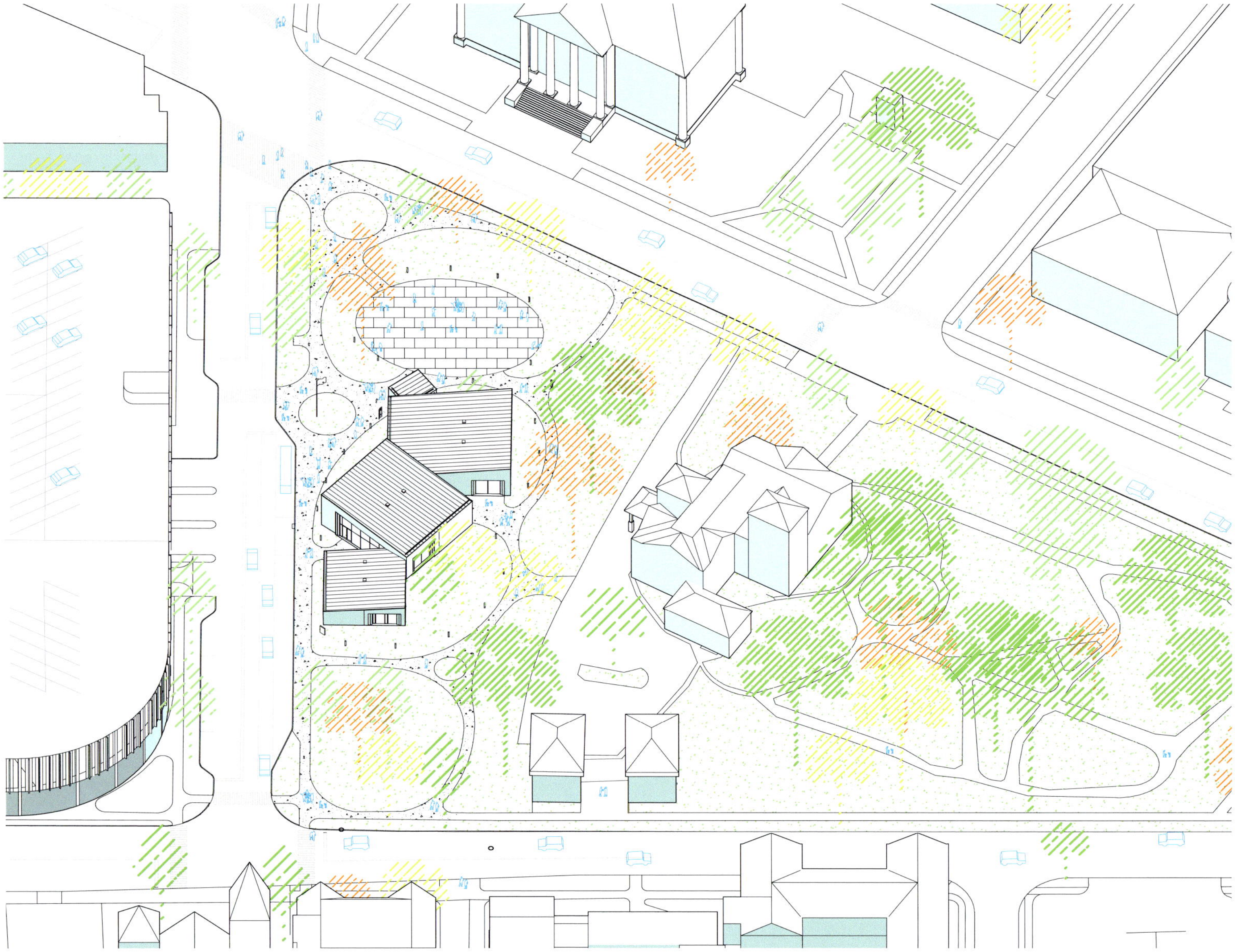

Axonometric plan

Floor plan

1

Billerica Memorial High School
Perkins&Will

Location Billerica, Massachusetts
Function High School
Client Town of Billerica, Massachusetts School Building Authority
Completion date 2020
Architect Perkins&Will
Sustainability Targeting LEED Silver Certification
Photography James Ewing

The town of Billerica has undergone a subtle yet significant transformation in recent years. However, its former high school, characterized by outdated and dysfunctional facilities, failed to accurately represent the town's contemporary identity. In collaboration with the community, Perkins&Will embarked on a project to reshape this perception by designing their largest civic structure. The newly constructed school accommodates over 1,600 students in grades 8-12, a Pre-K program, and district administrative offices.

Brick tectonics The architectural choice of brick serves as a symbolic lesson to students about the power of simplicity in generating complexity, as well as the potential for collective renewal. The masonry's interaction with natural elements visually narrates the ever-evolving nature of education's impact on our sense of self.

Drawing inspiration from the town's heritage as a New England mill town with a prevalence of red brick structures, the new high school seamlessly integrates this historical context with its forward-looking educational mission. The unassuming red brick exterior receives a fresh interpretation through geometric corbeling, while the interior employs sustainable timber construction techniques that infuse new significance into this traditional material. These simple

1 The school's neoclassical civic neighbors and late-1800s industrial context informed a design that connects distinct elements to speak a humble material language.

2 The building's elongated brick façade is organized by a repetitive frame expression.

3 Corbeled brick on the south façade takes measure of the day's light, amplifying shadows and quietly functioning as a masonry sun dial, tracking the path of the sun.

4 The entire interior uses daylight to capture the rhythms of academic time as the sun pushes shadow through the building.

5 Timber takes new meaning through carbon-conscious detailing; The embodied carbon in the FSC Certified spruce timber structure offsets the equivalent of a typical school bus traveling over 460,000 miles.

2

3

The masonry's interaction with natural elements visually narrates the ever-evolving nature of education's impact on our sense of self.

yet genuine materials are skillfully composed in three dimensions, conveying the school's historical narrative through an architectural language characterized by both introspection and projection.

The design approach takes into account the school's neoclassical civic surroundings and the industrial backdrop of the late 1800s. The resultant design embraces structural expressionism, wherein the connections between distinct architectural elements communicate a consciously modest material vocabulary. Echoes of historical styles pay homage to local landmarks like the Mill Buildings in North Billerica and the brick structures in the Town Center.

4

Strategically positioned, the building is oriented to maximize its presence along the primary route from the town. The design prioritizes communal spaces, ensuring that masonry facades shape the pedestrian experience, providing a graceful welcome to students, faculty, and visitors. The elongated brick façade, characterized by repetitive framing, incorporates windows of various sizes and functionalities through skillful use of stepping, corbelling, and coursing techniques. This dynamic façade serves as a metaphor for the 'maker' classrooms integral to the school's curriculum.

Luminosity The architectural approach transcends basic illumination needs by embracing natural daylight over artificial lighting. Both the interior and exterior utilize daylight to punctuate the rhythm of a student's day. Furthermore, the masonry's depth serves a practical purpose, offering solar shading to mitigate interior glare. Thus, these meticulously crafted brick surfaces simultaneously serve functional and aesthetic purposes, capturing the temporal cadence of academic life as the sun casts shifting shadows across the façade throughout the day, semester, and year.

Brick, a material forged from the earth, solidified by fire, and brought to life by the sun, encapsulates the delicate equilibrium between creativity and pragmatism that defines the building's essence. The process of bricklaying, involving intricate transformations, unveils a tapestry of intricate patterns and forms, capturing remnants of winter's touch and tracing the sun's ceaseless journey across the sky.

Both the interior and exterior utilize daylight to punctuate the rhythm of a student's day.

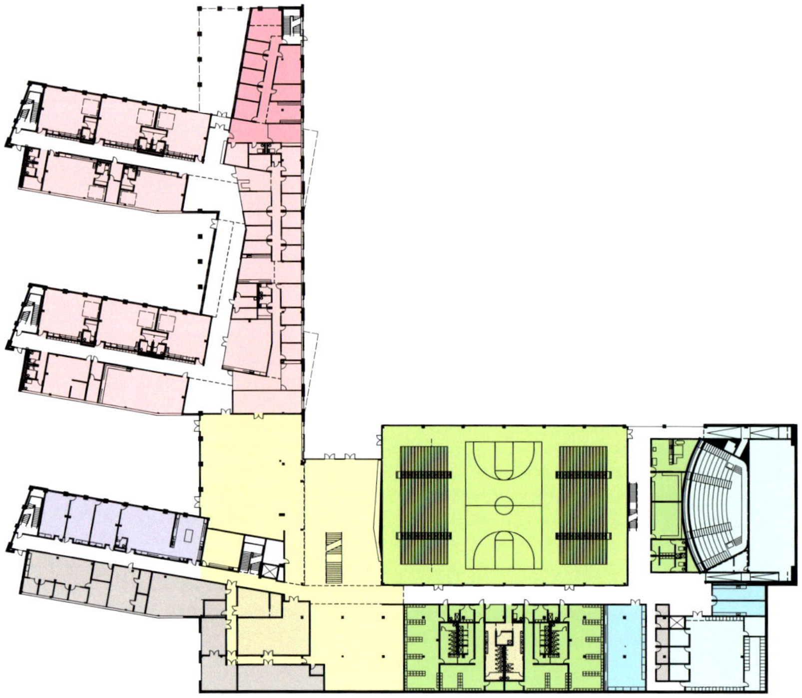

Level one

Level two

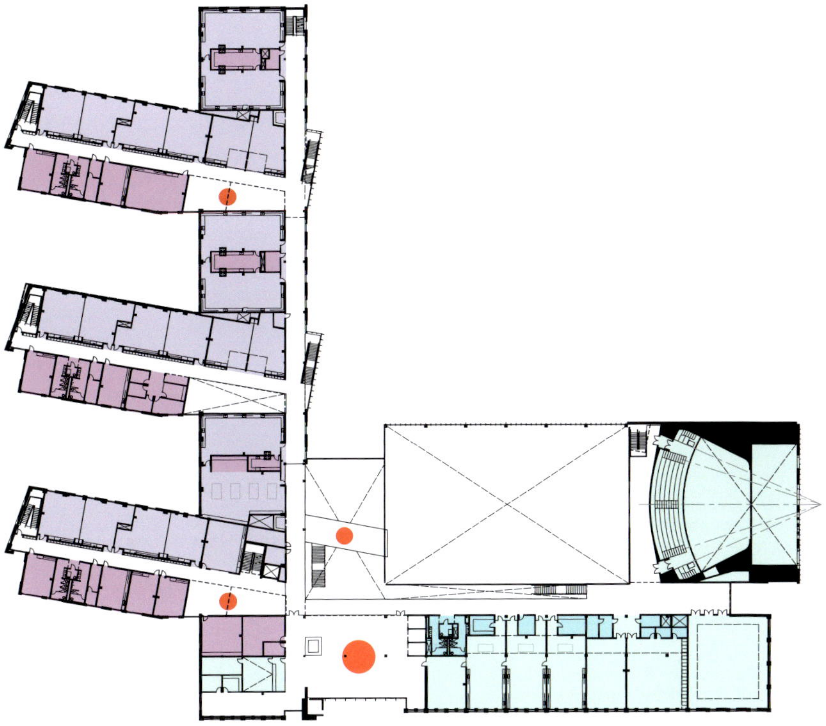

Level three

Level four

District office
Pre-K / Beam
Cafeteria
Cafeteria support
Service
Athletic
Athletic support
Art/Music
Art/Music support
Academic
Academic support
Media

Keene State College, Living and Learning Commons

Perkins&Will

Location Keene, New Hampshire
Function Living and learning commons
Client Keene State College
Completion date 2016
Architect Perkins&Will
Sustainability Carbon Neutral, ZNEB Ready
Photography Anton Grassi

1

The design incorporates distinctive detailing, infusing the façade with varied textures and reliefs, artfully modulating the building's scale.

2

3

The inception of the Living and Learning Commons (LLC) at Keene State College was driven by the aim to facilitate a seamless transition into college life for students. This objective was realized through the establishment of an environment in which both faculty and residential staff collaborate to foster communities that not only help students cultivate friendships but also thrive academically. Currently serving as the dedicated abode for 350 first-year students, the new LLC accommodates eleven diverse communities, each providing a range of opportunities for students to explore their academic pursuits and engage in co-curricular interests within a secure, supportive, and dynamic milieu.

Design Philosophy Strategically situated at the convergence of the east visitor entry and the Appian Way—a pivotal north-south axis connecting to the college's academic core—the LLC's design comprises three residential wings. These wings, clad in the iconic red brick reminiscent of the campus's heritage, delineate three distinct landscape areas. Complementing these brick residential wings is a vertical social hub, sheathed in metal and glass, serving as the connective thread of the community at the building scale. This central hub not only fosters engagement but also stands as a prominent landmark on campus, encouraging physical activity and interaction. This core component is set atop a base crafted from dark brick, firmly grounding the structure and housing essential spaces like classrooms, kitchens, and lounges, where students, faculty, and staff can mingle. Furthermore, at the terminus of each residential wing, the masonry façade gracefully parts to unveil serene study rooms overlooking the landscaped areas.

The exterior materials of the LLC harmoniously align with the college's overarching vision of a campus marked by material coherence. While adhering to the college's standard brick, the design incorporates distinctive detailing, infusing the façade with varied textures and reliefs, artfully modulating the building's scale. Complementing the brick, warm gray metal and glass are strategically employed to define public spaces, promoting transparency while offering a striking visual contrast to the sturdiness of the brick wings.

Sustainable stewardship Notably, the LLC emerges as a testament to environmental stewardship, meticulously designed to align with Keene State College's Climate Action Plan and the AIA 2030 Challenge. The building is engineered to drastically reduce carbon emissions by an impressive 70%, eschewing reliance on fossil fuels. The surrounding landscape plays a pivotal role in this effort, featuring an active rain garden that effectively filters rainwater, alongside an expansive lawn housing 25 concealed geo-exchange wells. The comprehensive energy optimization efforts yield a remarkable 44 EUI reduction compared to the energy code baseline. This progressive approach to sustainability is projected to yield a notable 10-to-12-year payback period through the carefully selected systems and strategies employed in the building's design.

4

The sky-lit central atrium connects essential parts of the building and introduces natural light throughout.

6

Complementing the brick, warm gray metal and glass are strategically employed to define public spaces.

1 A vibrant social and academic hub, the project fosters community and harbors many miniature neighborhoods under one roof.

2 The Common Lawn is both a gathering space for the precinct and a performative landscape which filters rainwater and accommodates geoexchange wells.

3 Warm wood surfaces are employed in main lounge areas and a vibrant color highlight identifies each floor.

4 The interior color palette is characterized by natural tones that connect the spaces with the surrounding landscape.

5 Three primary zones buzz with activity at the ground level - the main building entry, the student lounge with a family-style kitchen, and game room.

6 The Hub is the heart of the building and is vertically unified by a stair that winds its way up through the multi-story spaces, putting living and learning on display.

7 The ground level acts as the mixer, where students, faculty, and staff are encouraged to connect academically and socially.

7

Pedro Arrupe, S.J. Hall

Sasaki

Location Washington, DC
Function Residential Housing
Client Georgetown University
Completion date 2016
Architect Sasaki
Sustainability LEED Gold
Landscape Sasaki
Structural Engineer SK&A Structural Engineers, PLLC (structural)
Civil Engineer Zach Chrisco
Photography Alan Karchmer

Georgetown University faced unique design challenges and specific constraints when planning the Pedro Arrupe, S.J. Hall as part of a broader review of its student housing. The goal was to create a space that redefined how students live and learn while fostering campus engagement.

One significant challenge was finding the right location for new construction within the university's limited land resources. They chose a smaller transitional site, which, despite its small footprint, allowed for higher density due to its proximity to existing 5 to 8-story buildings. This location also helped integrate the new hall into the neighborhood, adding diversity to residential options and reinforcing a sense of community.

In the northeast quadrant, there was a lack of public space connecting student life. The project addressed this by creating a plaza that visually and physically links the new hall and the campus center. Previously, this area served as a circulation corridor, but the redesign maximized the site's

potential by placing the building's core against a hill and opening up public areas to the new plaza. This approach caters to various space needs, enhancing the building's functionality and livability within an academic context.

Environmental considerations Sustainability was a key consideration. The building and landscape design improved the site's ecological function by preserving mature trees, incorporating a green roof, using previous plazas to manage stormwater, and implementing native biocells along the perimeter. A large cistern reduced stormwater runoff and provided irrigation water, significantly lowering potable water demand. The project aimed for LEED Gold Certification and set new landscape material standards while complying with updated stormwater regulations at the University.

Materiality Regarding materials, the design aimed to blend tradition with contemporary identity. Stone was prominently used on the main tower facing the historic campus and at the base, allowing for spacious interior learning areas. The upper façades featured brick to match the surrounding structures, with subtle accents of cast stone. The use of large, operable windows not only invited natural light but also supported sustainability goals like natural ventilation.

3

The goal was to create a space that redefined how students live and learn while fostering campus engagement.

This location also helped integrate the new hall into the neighborhood, adding diversity to residential options and reinforcing a sense of community.

4

5

1 The student lounges are highly visible and telegraph student activity day and night from the new plaza and pedestrian path.

2 The 225-bed Pedro Arrupe, S.J. Hall is a new paradigm for residential life on campus.

3 Porous indoor/outdoor environments along the new plaza and pedestrian way ensure that the inner life of the living and learning hub is visible to the broader campus and inviting to the student community.

4 The main tower, which faces the historical campus, is clad in stone.

5 The first floor hub is a direct recognition that learning happens throughout campus, at all hours. The flexible studio is equipped to accommodate everything from a traditional class to a cross institution conference.

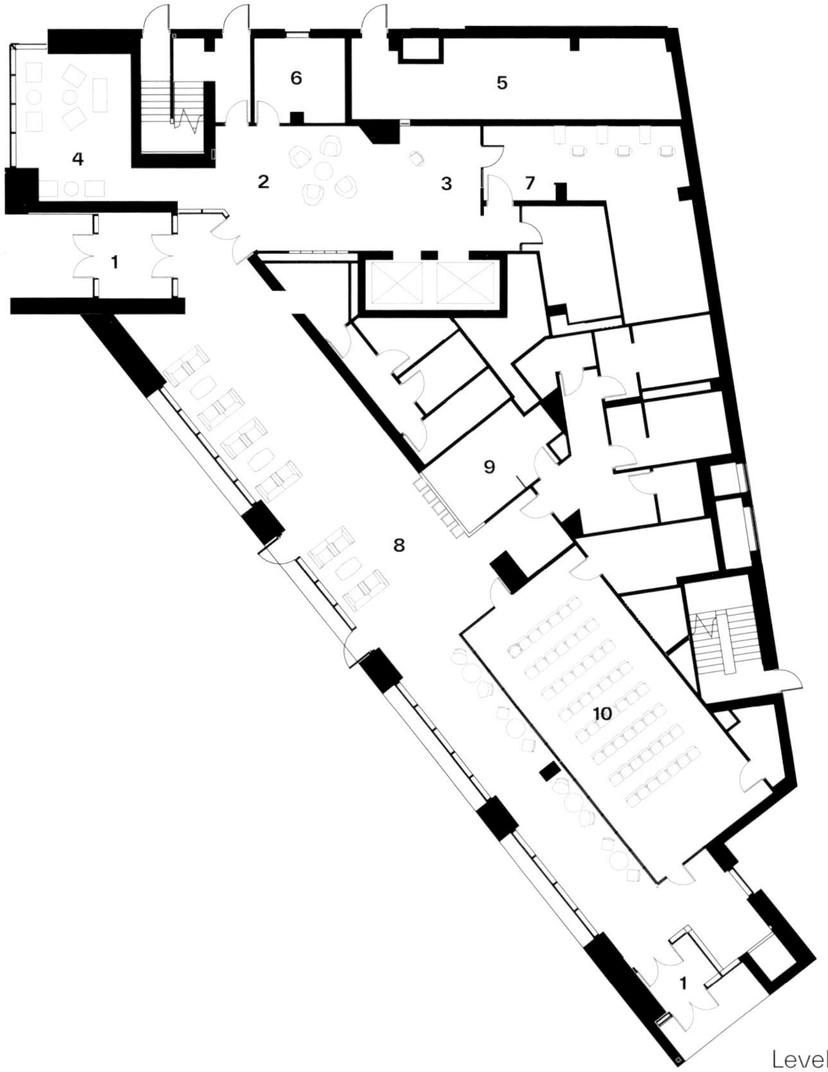

Level 1

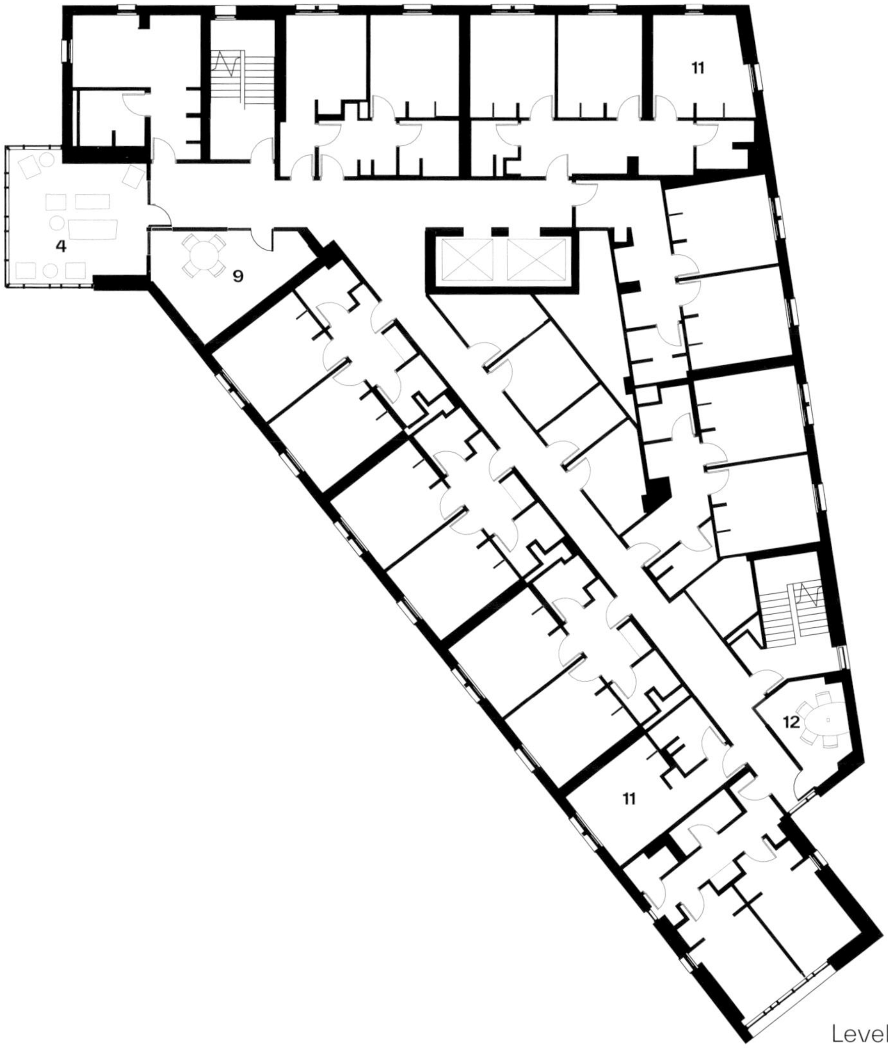

Level 2

1 Vestibule
2 Lobby
3 Control Desk
4 Lounge
5 Bike Storage
6 RD Office
7 RHO Office
8 Workroom
9 Kitchen
10 Multi Purpose Room
11 Bedroom
12 Study Room

1

Singleton, Hitch, and Maida Residential Houses

Sasaki

Location Bethlehem, Pennsylvania
Function Residential Houses
Client Lehigh University
Completion date 2020
Architect Sasaki
Sustainability LEED Silver certification
Landscape Sasaki
Engineer LeMessurier
MEP/FP Engineering Vanderweil
Engineer Civil Langan
General Contractor Whiting-Turner Contracting
Photography Ty Cole

The Lehigh University Singleton, Hitch, and Maida Residential Houses initiative embodies our commitment to elevating the residential experience in line with the University's comprehensive enhancement plan. This innovative project, conceived to accommodate the institution's growth while ensuring parity in student housing, has yielded remarkable outcomes.

The recently established residential houses offer a departure from conventional dormitories, emphasizing increased privacy while preserving a strong sense of community. Through a thoughtful fusion of academic and social spaces, these facilities encompass a fitness center, café, lounges, versatile multipurpose areas, and communal kitchens. These provisions are pivotal in nurturing the overall well-being of students engaged in demanding academic pursuits.

Contextual considerations The distinct topography of Lehigh University, characterized by steep inclines, posed challenges in creating north-south connections. This prompted the campus layout to predominantly revolve around east-west terraces. To address this topographical complexity, the design unfolds as a series of cascading experiences tailored to students. By introducing new entrance courts and terraces that traverse the undulating terrain, the design harmonizes with the organic layout of the campus' open spaces.

The landscape architecture seamlessly integrates the residential hillside with the university's historic core through the establishment of flat, open spaces—an invaluable asset within the campus. With a meticulous assessment of the surrounding woodland ecology, the project's team refined their planting strategy, enriching the core campus landscape and promoting habitat restoration.

Inside the newly erected structures, a coherent sequence of interconnected spaces dedicated to health, well-being, study, and social interaction seamlessly interlink both horizontally and vertically. This arrangement facilitates effortless movement within the residential houses and towards the broader campus precincts. The design strategy encompasses both campus-wide and district-specific considerations, creating a contextual resonance, both in

The design unfolds as a series of cascading experiences tailored to students.

2

3

terms of physical layout and programming. The innovative design establishes a connection between the historic stone buildings dating back to the 19th century and the upper-campus residential district.

Thoughtful materiality The buildings are crafted with materials that blend harmoniously with the surrounding architecture, sculpted to align with traditional residence halls. The selection of materials, featuring diverse brick textures and metal cladding, infuses dynamism into the building forms, while the play of light adds depth. The strategic use of glazed elements fosters luminosity and establishes a seamless visual link between indoor and outdoor spaces.

The buildings are crafted with materials that blend harmoniously with the surrounding architecture.

4

1 The exterior material palette expresses a modern take on traditional masonry, and the buildings establish their own bold identity while harmonizing with the existing campus context.

2 The design is conceived as a series of cascading moments for student experience. New courtyards and terraces offer a complement to the organic nature of the campus' open spaces.

3 As an important component of the campus' social and residential life, the project intentionally integrates living, wellness, study, and social space.

4 The project amenities cater to diverse needs and provide a sense of community among residents by encouraging human connection and fostering socialization and relationships.

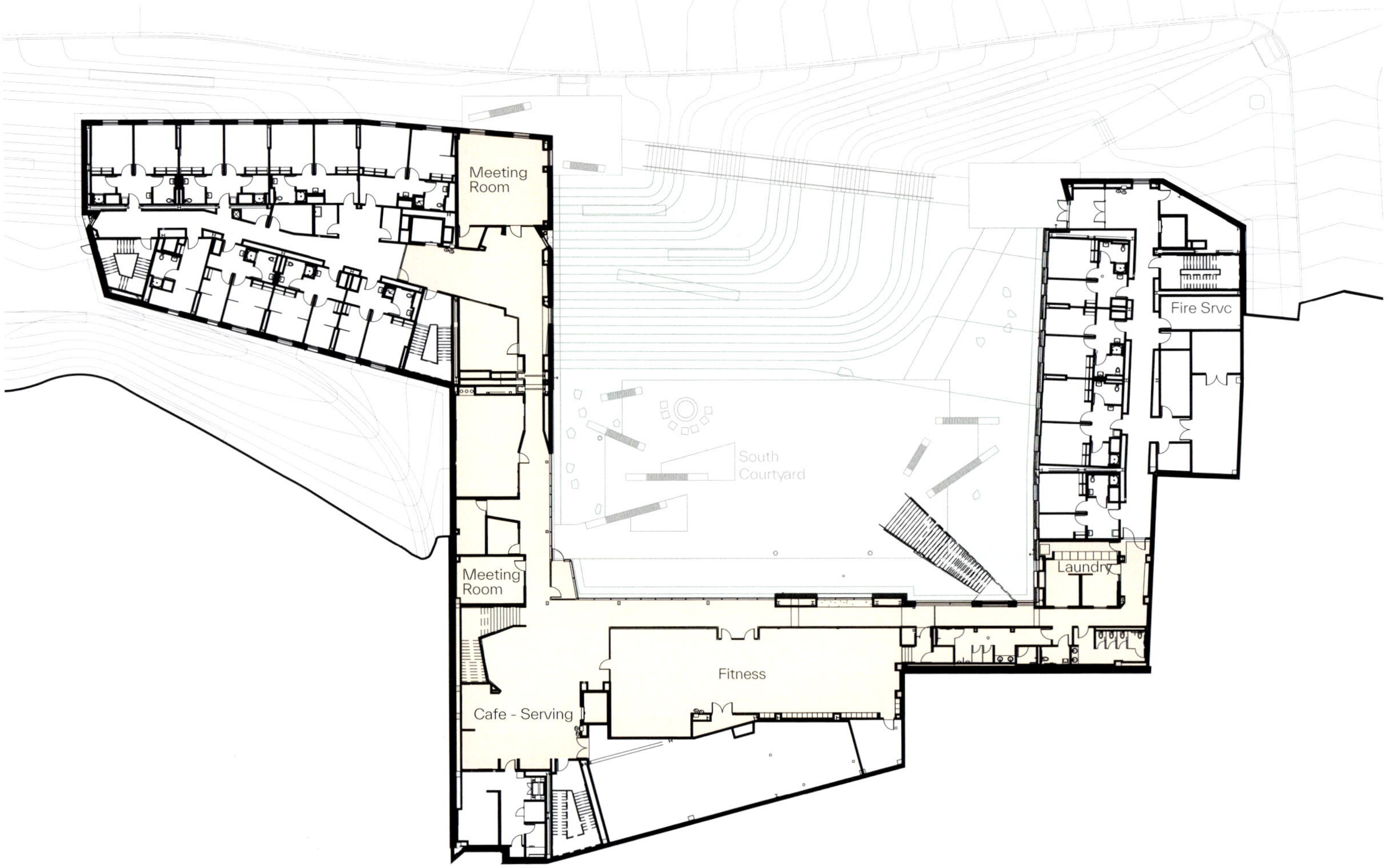

Level three plan

Level four plan

Pauli Murray College and Benjamin Franklin College

Robert A.M. Stern Architects

Location New Haven, Connecticut
Function Residential Colleges
Client Yale University
Completion date 2017
Architect Robert A.M. Stern Architects
Sustainability LEED Gold
Landscape Olin Partnership
Civil Engineer Tighe & Bond
Contractor Dimeo Construction Company
Lighting Buro Happold Consulting Engineers, P.C.
Photography Peter Aaron/OTTO

Yale's system of residential colleges is the cornerstone of its undergraduate experience. Between the two world wars, under the leadership of James Gamble Rogers (1867–1947), ten residential colleges were built, four employing the red-brick Georgian language, but six, as well as many academic buildings, adhering to the Collegiate Gothic style with which Yale is most identified today. Whether Georgian or Gothic, the colleges were organized as quadrangles around variously sized courtyards, and in their totality achieved a finely calibrated urbanism comparable to that of England's Oxford and Cambridge universities.

Two additional colleges, designed by Eero Saarinen and completed in 1962 with less success, honor Rogers's urbanism but in a distinctly modern way. As we set out to design the first new residential colleges to be built since Saarinen's, we fully embraced the Rogers model, allowing it to show us the way, not only with his expressive interpretation of the Gothic but also with the genius of his urbanism.

The design of these two new colleges, housing 450 students each on a triangular 6.2-acre site bisected by a landscaped

1

Conceived as fraternal twins, similar in size but each enjoying its own identity.

3

4

1 Pauli Murray College Bekenstein Court.

2 Prospect Walk.

3 Benjamin Franklin College Class of 1964 Gate.

4 Benjamin Franklin College brickwork detail.

5 Pauli Murray College brickwork detail.

6 Benjamin Franklin College Nyburg Baker Court.

5

public pedestrian way, are conceived as fraternal twins, similar in size but each enjoying its own identity, each incorporating a dining hall, a library, and a house for the head of the college, and each maintaining the traditional organization of entryways that intentionally create more intimate communities of students within the larger whole.

Echoing Rogers's environmentally responsive massing, building heights rise from south to north to allow for maximum sunlight in the courtyards. The new colleges add to Yale's lively skyline with a variety of dramatically modeled towers, visually collapsing the distance between Yale's Central Campus and the precinct of Science Hill to the north. Taken as a whole, the two new colleges reflect our conviction that physical form is essential to fostering community.

The two new colleges reflect our conviction that physical form is essential to fostering community.

6

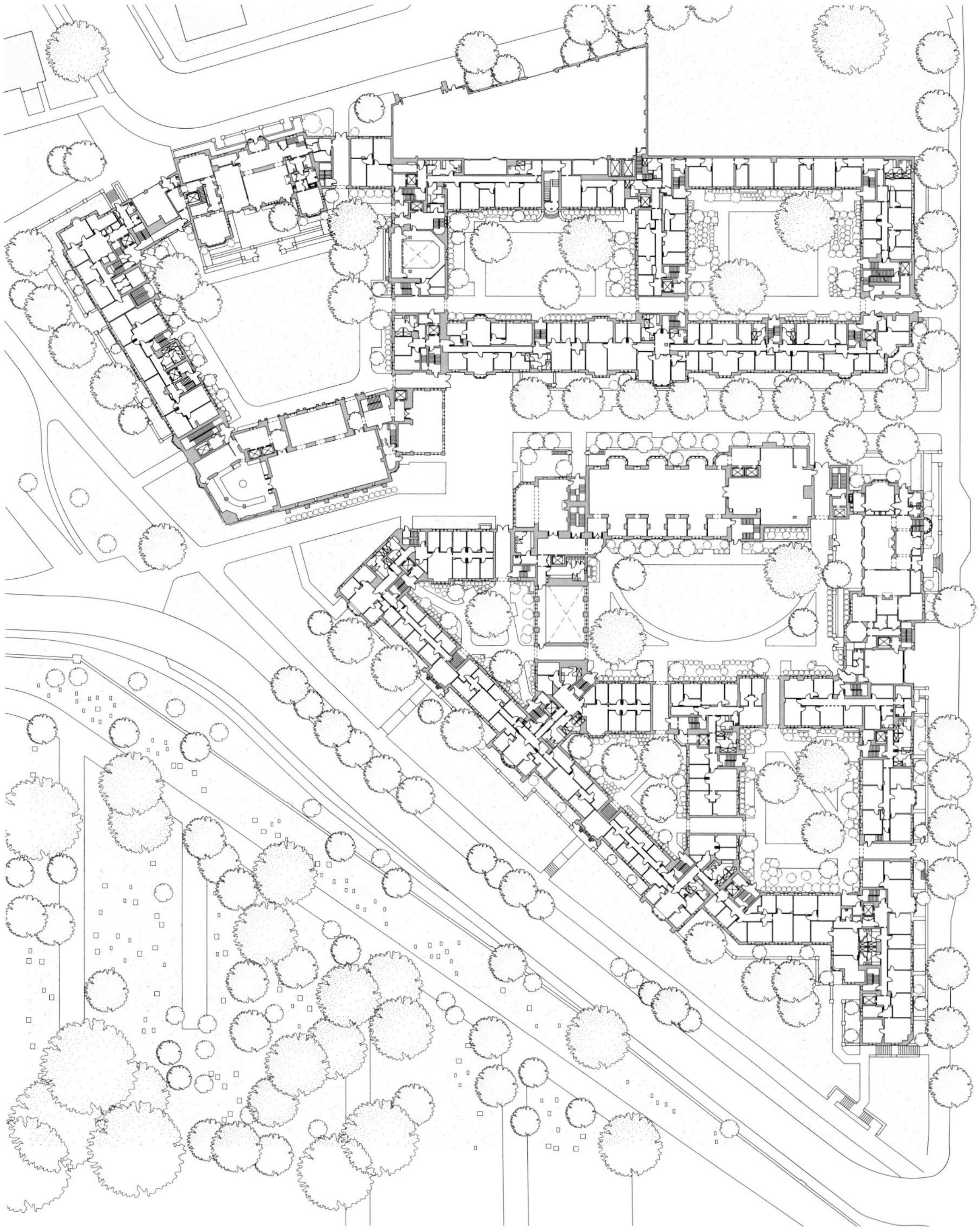

Ground floor plan

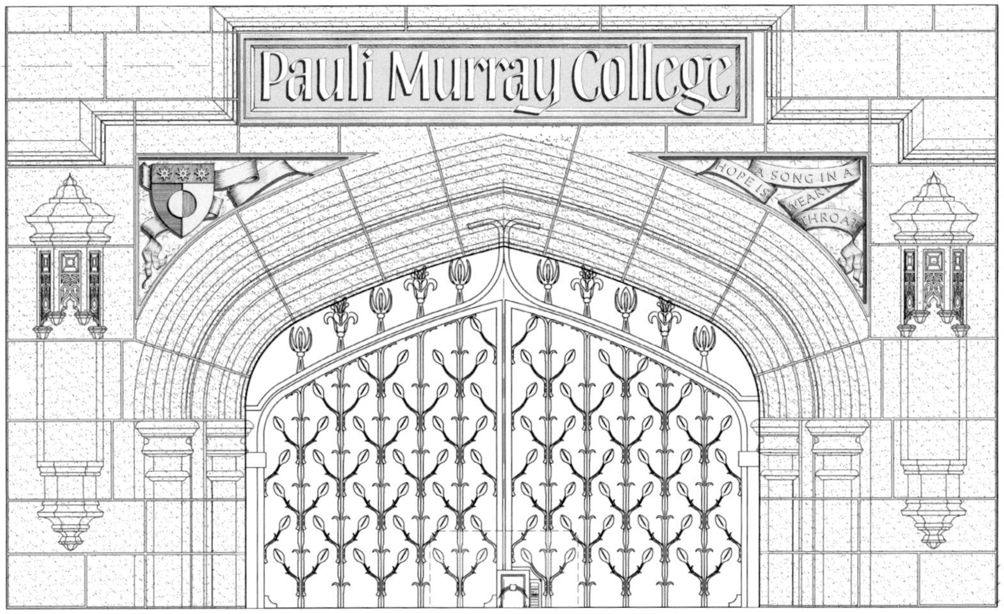

Pauli Murray College archway

Benjamin Franklin College archway

Steel Plant Studios

Robert A.M. Stern Architects

Location Poughkeepsie, New York
Function Arts Hub
Client Marist College
Completion date 2019
Architect Robert A.M. Stern Architects
Landscape Imbiano Quigley Landscape Architects, PC
Structural & Envelope Engineer Thornton Tomasetti Engineers
MEP Engineer Kohler Ronan
Civil Engineer By Client
Photography Peter Aaron/OTTO

The building's versatility and advanced infrastructure are designed to address both current and future students' needs.

2

3

The Steel Plant Studios, an innovative addition to Marist College, serves as a hub for Fine Arts, Digital Media, and Fashion programs, known for their global recognition. This state-of-the-art building combines a refurbished 12,000-square-foot historical industrial structure with a 35,000-square-foot contemporary extension featuring bright, adaptable studios and classrooms. The Steel Plant Studios not only enhances Marist's campus layout but also acts as a cultural focal point for the growing East Campus.

Drawing inspiration from the original building's industrial aesthetics, the new L-shaped extension maintains its legacy through large factory-style windows and locally sourced, hand-crafted brick cladding. Glass-enclosed sections extend from the primary brick structure at the southwest and northeast corners, covered by patterned aluminum screens that bear a textile-like representation of the Marist order's emblem.

The Steel Plant Studios accommodates Marist's expanding curriculum, which centers on the practical application of knowledge. It incorporates professional-grade maker-spaces, digital pattern-making laboratories, and a novel retail area. One noteworthy feature is the M-Porium, a functional storefront where students can gain hands-on experience in exhibiting, marketing, and selling their creations. The classrooms and studios are equipped with cutting-edge technologies, and adaptable sliding walls allow for the creation of two 70-foot long fashion runways. The building's versatility and advanced infrastructure are designed to address both current and future students' needs.

A central highlight of the new structure is the Winter Garden, nestled between the old and new portions. This courtyard-like space has transformed into the heart of the building, and by extension, East Campus, fostering a sense of belonging and community. The architectural blend of modern and industrial elements gives the impression that the extension developed organically over time, symbolizing the college's interdisciplinary approach.

Marist College's Steel Plant Studios stand as a testament to its commitment to holistic education. By providing world-class learning environments and bridging the campus divide, this innovative addition cultivates creativity and a sense of unity among students. The Steel Plant Studios' design, blending history with modernity, exemplifies Marist's dedication to preparing students for the challenges of the twenty-first century global society.

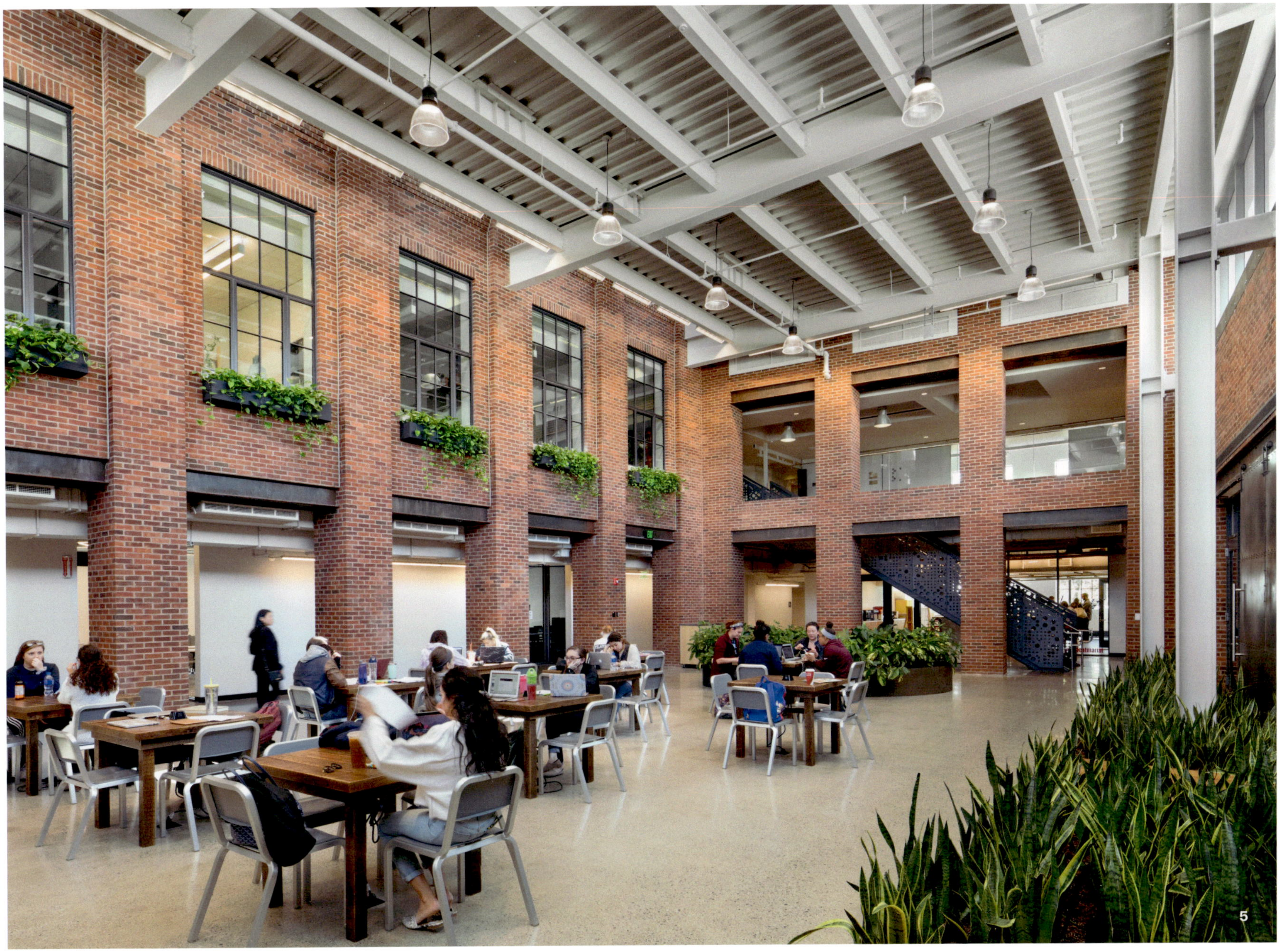

The Steel Plant
Studios' design, blending history with modernity.

1 View of main entry.
2 View from Route 9.
3 Winter garden detail.
4 Stair.
5 Winter garden.

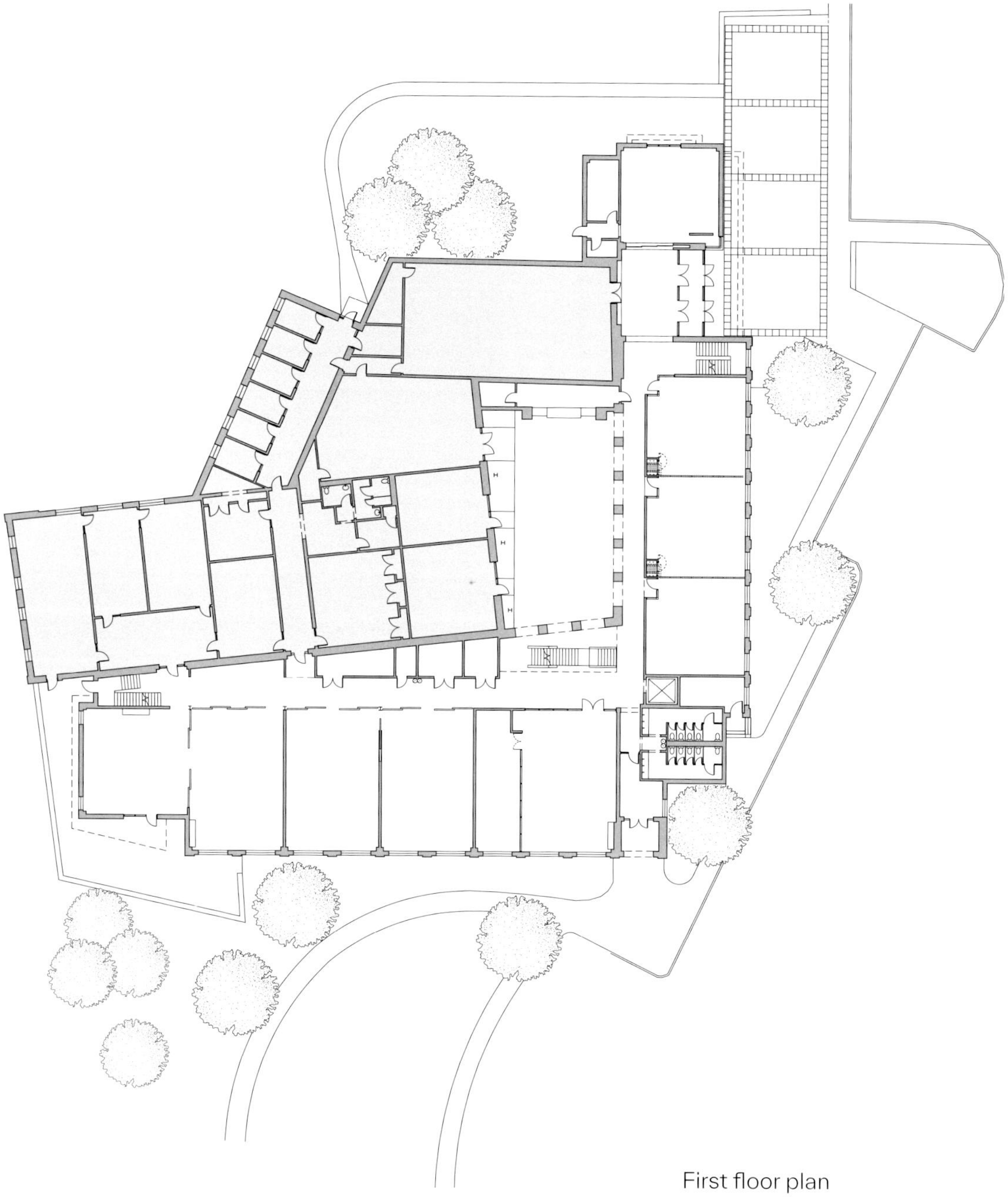

First floor plan

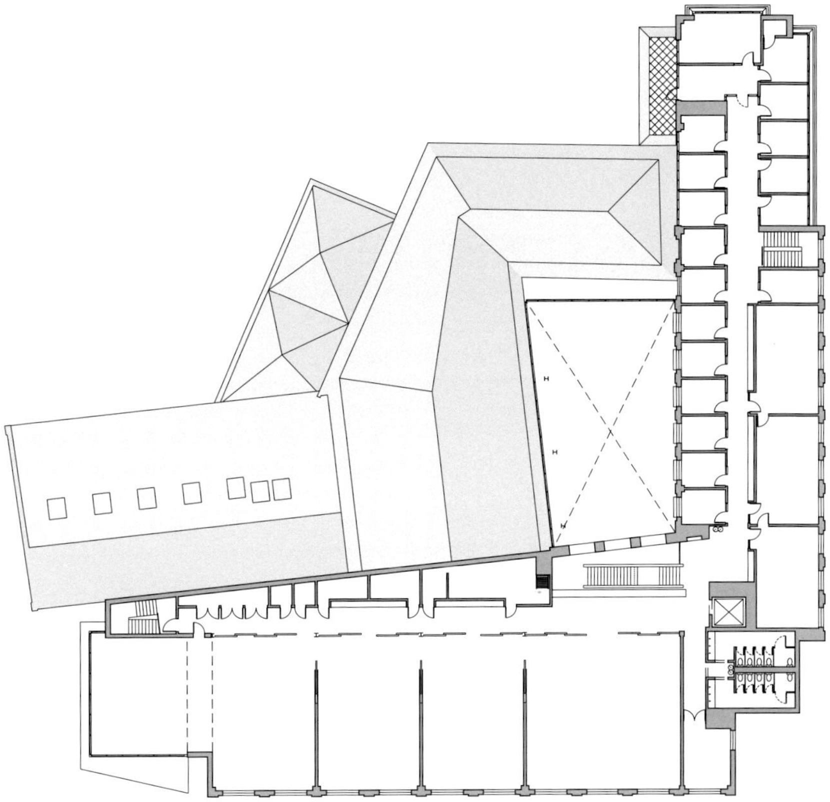

Second floor plan

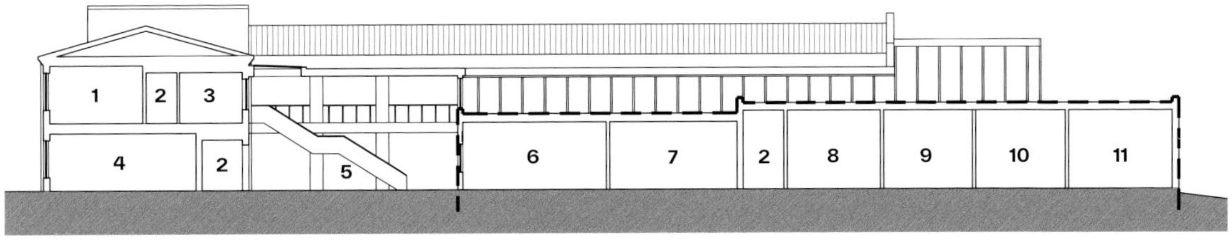

Section

1 Shared lab
2 Hall
3 Office
4 Classroom
5 Winter garden
6 Trend lab
7 Photo studio
8 Arts digital lab
9 2D Fundamentals
10 Print studio
11 Sculpture studio
– Dashed line indicates existing building line

Center for Jewish Life

Stanley Saitowitz | Natoma Architects Inc.

Location Philadelphia, Pennsylvania
Function Educational Center
Client Drexel University
Completion date 2016
Architect Stanley Saitowitz | Natoma Architects Inc.
Sustainability Green Building Initiative building
Landscape Bohler Engineering
General Contractor Bittenbender Construction LP
Structural Engineer CVM
Photography Richard Barnes

Drexel University's Hillel House is located on the university's campus across the Schuylkill River from Downtown Philadelphia. The building takes its cues from the city's architectural heritage and is sheathed in local red brick as textured fabric draped in an abstract menorah that terraces down to the street. The façade design references the branched lampstands used in Jewish homes and temples, as well as the striated garments worn during prayer.

Arranged on four interconnected levels, the square building has thickened side walls which contain services, and four central columns which structure the middle, front and rear.

Support (services) The basement contains the kitchens, storage and mechanical areas.

Action (community) The street level is dedicated to everyday life, to meeting and talking, sitting around a fire, gathering and eating at the rear which opens to a garden. At the center, and connecting the second floor, is a stair, amphitheater, auditorium, with a balcony overlooking it from above. This multilevel courtyard is a multipurpose space at the heart of the building, the pivot of everything.

1

The façade design references the branched lampstands used in Jewish homes and temples, as well as the striated garments worn during prayer.

1 Front façade sheathed in local brick.

2 Brickwork façade referencing a branched lampstand.

3 Worship (divinity) floor.

4 Brick façade mimicking the surround context.

5 View through the exterior courtyard.

6 View down the amphitheater.

5

Formation (learning) The second floor is for focused activities, offices for planning and arranging, rooms for quiet study and discussion, places for groups and gathering, spaces for exploration and learning.

Worship (divinity) The top floor is the place of worship, the Shabbat floor, the level of sanctity. The three prayer gatherings, conservative, orthodox and reform, are connected with a central court that opens with a circular cut, an absent dome to the sky above. This is the only view in the city free of the works of man. This absent sanctuary is what is common and connects the three branches of worship in shared faith. The largest room doubles as the library, or Beit Midrash, the House of Learning.

By day, light reflects on the woven brick notches, by night light shines from within. This is a space to create a continuing community of traditional values through meeting, learning, and ceremony. This is a place to gather and emanate light.

This is a space to create a continuing community of traditional values through meeting, learning, and ceremony.

6

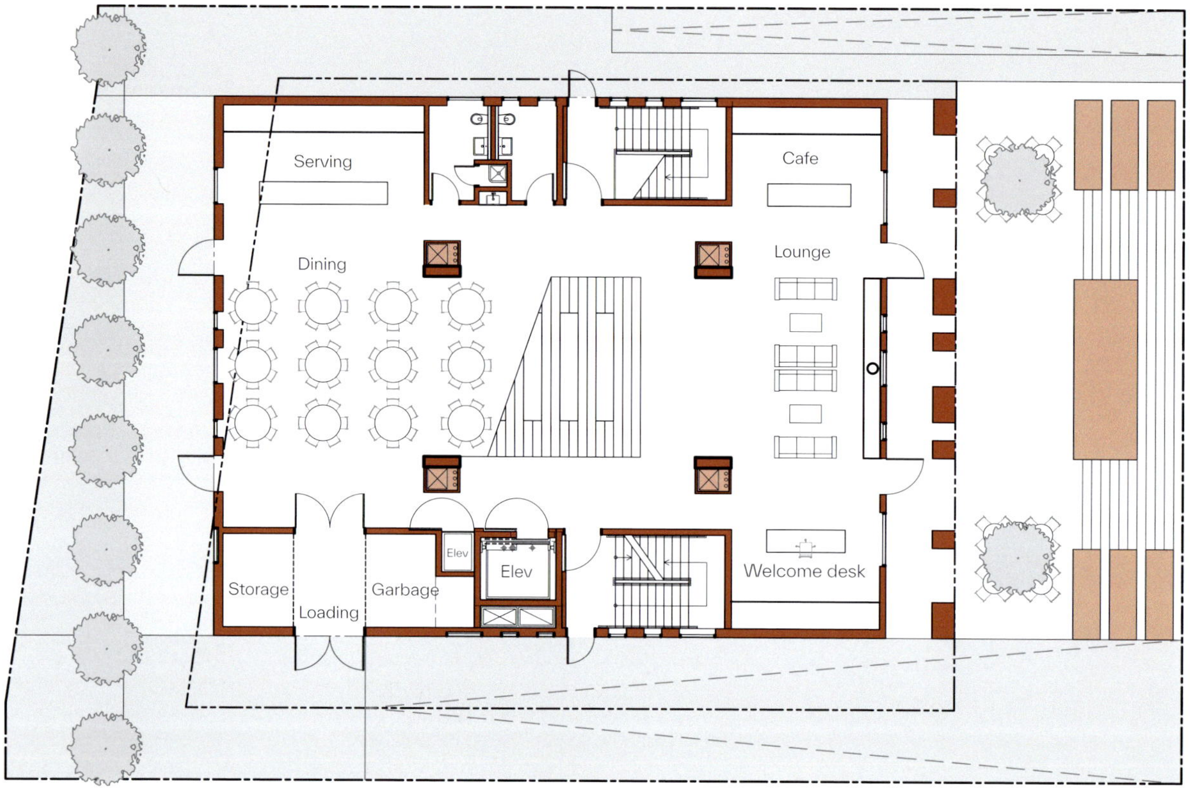

Ground floor plan

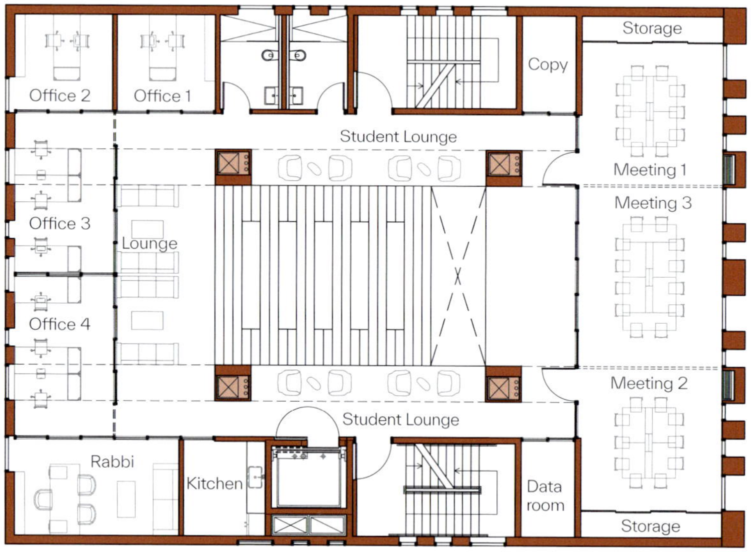

Second floor plan

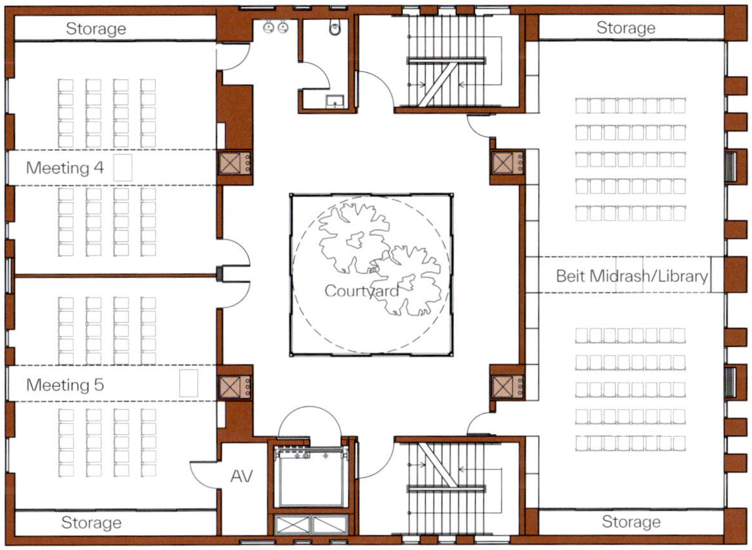

Third floor plan

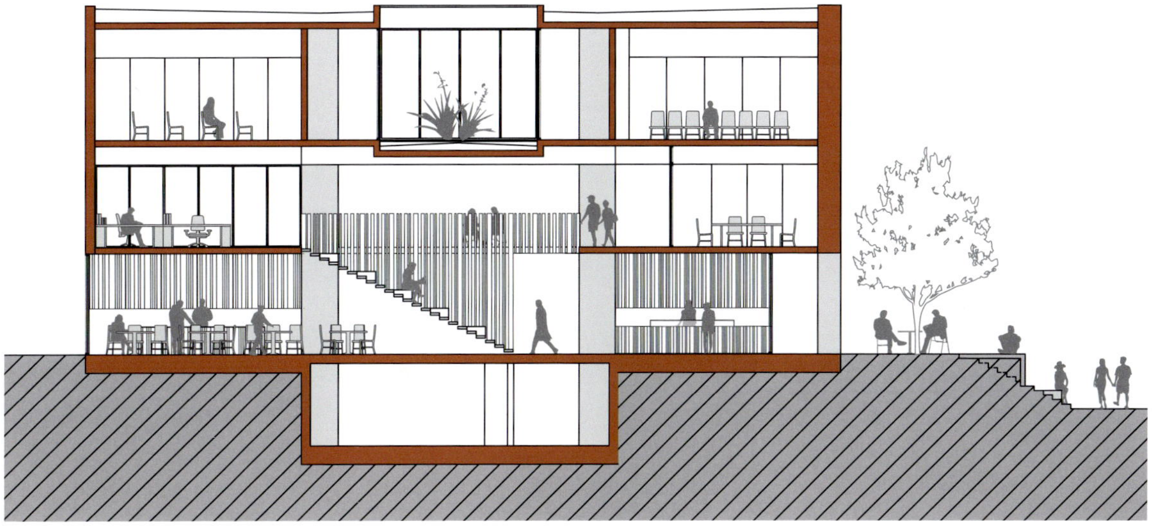

Section

1

MTSU Student Services Building

Tuck-Hinton Architects

Location Murfreesboro, Tennessee
Function Student Services Building and Garage
Client Middle Tennessee State University
Completion date 2013
Architect Tuck-Hinton Architects (now Anecdote)
Sustainability State of Tennessee, High Performance Building Requirements
Landscape Architect HDLA
Civil Engineer Barge Cauthen Associates
Structural Engineer EMC Structural Engineers
MEP I.C. Thomasson Associates
Contractor Messer Construction Company
Photography Thomas K. Gatlin

In 2010, Tuck-Hinton Architects was commissioned to design Middle Tennessee State University's Student Services Building and Garage. The site is on the east side of its campus located in Murfreesboro, Tennessee.

The building's 60,000 square foot program includes a Bursar's Office, Business Office, Admissions, Financial Aid, Records/Scheduling, Enrollment Services, ID/IT, a tour room for orientation, conference and support spaces, and the university's first parking structure.

Underpinning the design brief was the need to create a space that could welcome, support, and guide students at one of the state's largest university.

One of the two 500 car garages is positioned to make the Student Services Building the front door to the east campus entrance. An existing pedestrian corridor is extended from the west to a courtyard on the south side of the building. An iconic steel structure pedestrian bridge connects the garage to the building and passes through it, eventually crossing Blue Raider Drive and connecting to the existing Student Union. The bridge structure features infill panels of mesh and patterned glass.

The long, colonnaded east façade welcomes visitors, its large glass openings and multiple entrances providing views inward.

2

A linear composition positioned directly across the street from the massive Student Union, this two-level building has four distinctly different faces, each responding to the current and anticipated character of the growing institution.

The long, colonnaded east façade welcomes visitors, its large glass openings and multiple entrances providing views inward and convenient access from a nearby new campus entrance. The south colonnaded façade welcomes students traversing the school's main east-west pedestrian walkway, and together both façades provide shade and weather protection.

3

1 Curved brick at north façade.

2 Colonnade at east façade.

3 Student services building, bridge and parking garage

4 Interior view of south lobby.

5 View from upper corridor to lobby below.

6 South façade.

4

Underpinning the design brief was the need to create a space that could welcome, support, and guide students at the state's largest university.

5

In contrast, the west façade is a restrained cadence of multi-pane windows, while the curved north façade responds in form to the nearby campus roundabout. Circulation within this facility is intentionally straightforward, with an emphasis on accessibility to the important departments and services. The project includes the university's first parking garage, with an elevated and partially enclosed passage connecting it to, and through, the new building. As this corridor crosses above the main lobby, large graphics on the glazed surface celebrate the institution's promotional theme, 'True Blue'.

6

This two-level building has four distinctly different faces.

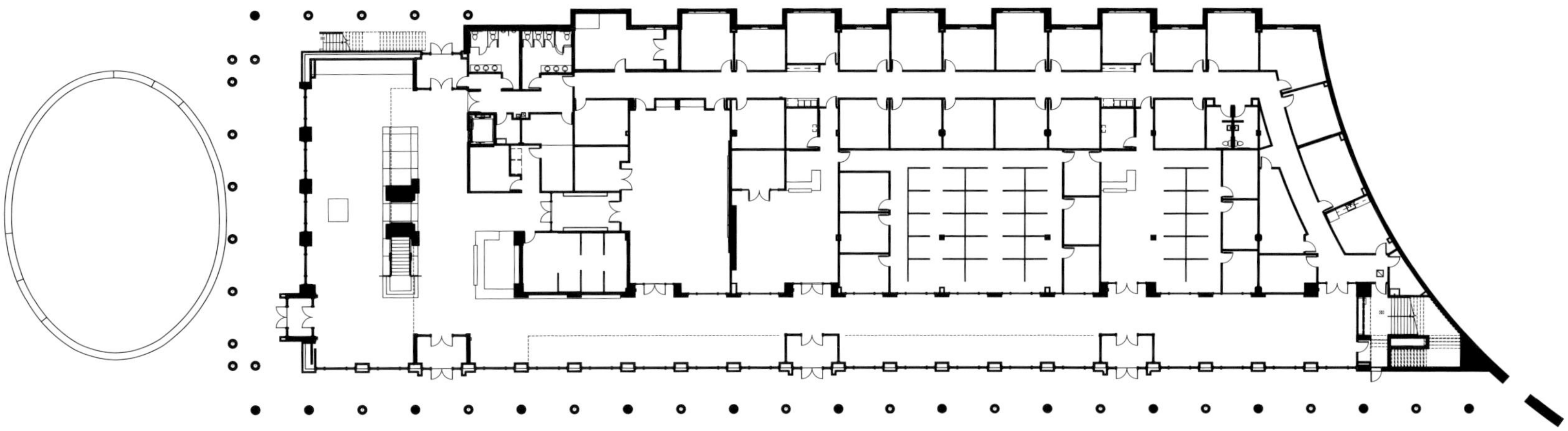

Level one plan

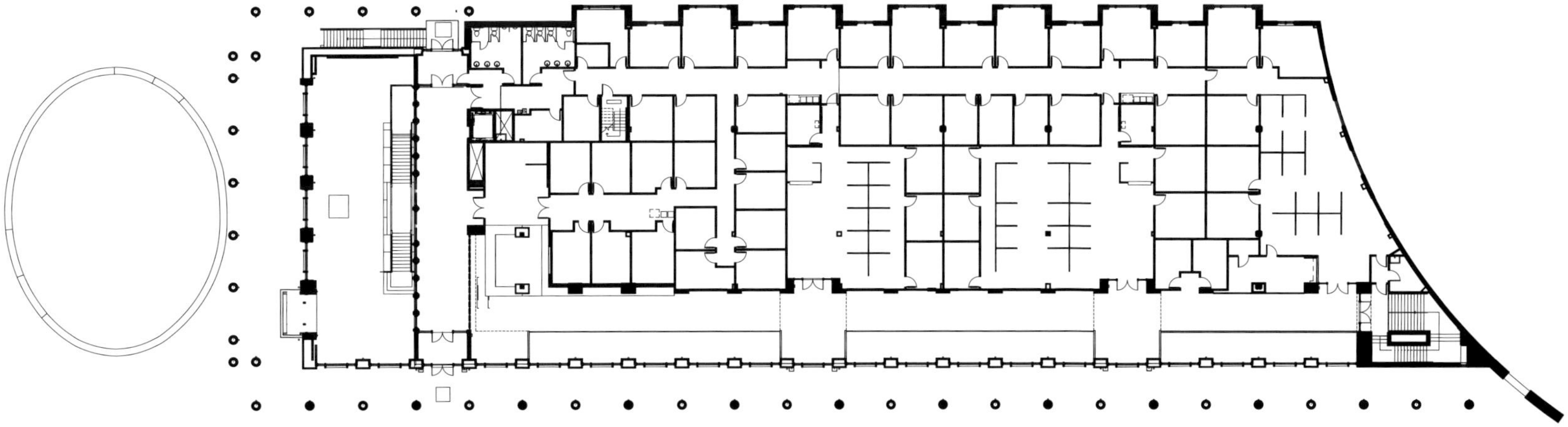

Level two plan

Beautiful brick
Illustrated studies

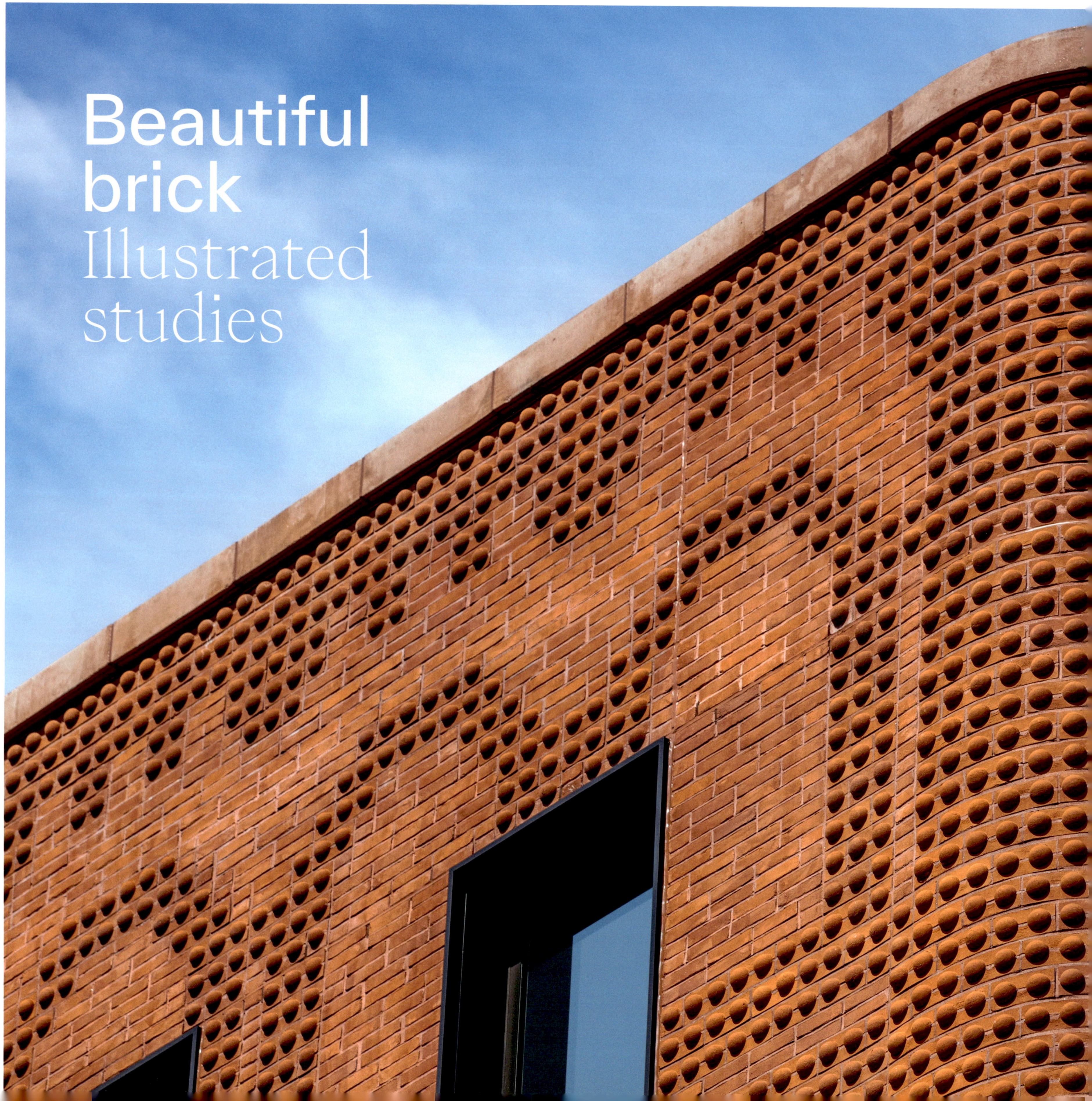

Consensus and connection: architecture of the authentic

Derek Marks & Steve Knight

con•sen•sus (kən-sĕn ' səs) n. [Lat. < *consentire*, to agree.—see CONSENT.] 1. Collective opinion. 2. General accord: AGREEMENT.

People generally know what they want, what works, and what doesn't in their built environment . . .

con•nec•tion (kə-nĕk ' shən) n. 1. The act of connecting or the state of being connected . . . 2. An association or relationship <a connection between the two incidents>. 3. Logical ordering of words or ideas : COHERENCE. 4. The relation of a word or idea to the surrounding text : CONTEXT <In this connection the wrong word was used>.

. . . and architecture is the artful combination of observation, critical listening, and skillful discernment to attain it.

au•then•tic (ô-thĕn ' tik) *adj*. [ME *autentik* , OFr. *autentique*, LLat. *authenticus*, Gk. *authentikos*, *authentēs*, author.] 1. a. Conforming to fact and therefore worthy of trust, reliance, or belief. b. Having an undisputed origin : GENUINE. 2. *Law*. Executed with due process of law . . . 4. *Obs*. Authoritative. —au•then ' ti•cal•ly *adv*.—au'then•tic ' i•ty (ô-thĕn ' tīs ' ə-tē) *n*.

For more than four decades, David M. Schwarz Architects has practiced architecture on a wide variety of building types across the United States and overseas. We design according to the conviction that authentic architecture comes from consensus and is shaped by connection. We seek consensus through a process of discovery and diplomacy in a team-oriented working environment. We study multiple approaches to layout, detail, architectural expression, and building materials. Ideas are vetted throughout the project team and the result is a response each team member can feel invested in. Connection is both intangible and tangible: we seek to create buildings that are connected to community aspirations and are connected to their physical, social, and programmatic context as well. We believe buildings can and should be monumental yet comfortable, exceptional yet familiar, and inspired by the past but not a caricature of it. The decisions we make together concerning a building's palette of materials are particularly important in this regard.

1

The human scale

Architecture, particularly in the public realm, is the one art that people have little to no choice to interact with. We can choose to look at a painting or not, listen to a piece of music or not, and watch a play or not. Rarely, however, can we exact such choices in the built environment. It is with this premise that we firmly hold the belief that buildings should be comfortable for the people who use them, whether on a regular active basis or on a passive one. A large building—a concert hall, arena, or courthouse—can be comfortable when a connection is struck between its components and the people who use it. Careful consideration and application of devices like materiality, ornament, and detail to scale large buildings to be approachable to people are key.

For instance, at Dickies Arena in Fort Worth, Texas, we crafted a blend of brick, stone, terrazzo, metal, and wood that is warm and welcoming. This palette and the level at which it is applied prioritizes human scale over big gestures.

2

We worked with the client to craft a building distinctively of north Texas. The brick, specifically, is a golden-hued blend, selected by architect and client after reviewing nearly 80 mock-ups on-site in varying lighting. It references many of the masonry buildings in Fort Worth including the neighboring Will Rogers Memorial Center. Other outcomes include expansive mosaic panels above the two primary entrances that tell the story of the American West and iconography drawn from native flora that is incorporated into carved stone panels, floor patterns, and railing details that connect the building to its context.

3

Form and function

Designed and executed using a skilled application of detail and material, places can be both exceptional in quality yet familiar in form and function. We have applied this philosophy to the design of four residential colleges at Vanderbilt University in Nashville, Tennessee. The residential college program was championed by the school's Chancellor, Nicholas S. Zeppos, who, in his words, believed that 'the colleges had to not only support and extend (Vanderbilt's) academic mission, but also provide places of connection and community: they needed to offer the full breadth of what Vanderbilt had to offer. They needed to make (Vanderbilt's) students' lives richer, their education more immersive and fulfilling'. The four brick and stone Collegiate Gothic colleges, which offer a program of public and resident spaces including a great room, dining hall, student activity room, academic space, and cloistered courtyards, are authentic to the Chancellor's ambitions. Furthermore, corridors on the student room floors are punctuated by nodes and offsets when they meet stairs, elevators, lounges, and studies to break down their scale and encourage interaction and community building among residents. Slight adjustments to the brick blends and patterning used on each of the four colleges subtly distinguish one from another, while maintaining a strong outward facing image of this institution to the broader community.

Precedent and process – tools to construct ideas

A deep understanding of precedent is the foundation of our design process. We believe that the study of precedent is quite different from just looking to and recreating places from the past. Rather, it is a tool to construct ideas that build consensus among stakeholders and connection to place, because immersion in the rich continuity of precedent means we are continuing the conversation with culture in the present and future, which we believe is foundational to an architecture of consensus. We often use one precedent as a springboard to others. This allows us to study varying approaches to form, function, and detail. A building's materiality is an important facet of precedent as well: many communities (geographic, historic or institutional) have long-established traditions of building with certain resources that need to be understood and included in this conversation.

4

A deep understanding of precedent is the foundation of our design process.

Using precedent to present contrasting approaches to similar design questions, whether related by material, program, formal typology, or aesthetic vernacular, stimulates dialogue and, ultimately, results in projects intrinsically tied to their place, purpose, and constituents. In addition to being durable by its very nature, brick figures prominently in the traditions of institutions and communities.

The process behind this approach requires patience and investment, but it yields highly relevant results for people and communities. Furthermore, the results prove to be worth holding on to and caring for. For the architect, client, and user, architecture of consensus and connection yields places that will last. A concert hall, civic building, or plaza, authentic to its place and purpose, will be embraced and used for generations—a true measure of sustainability.

Derek Marks and Steve Knight are architects at David M. Schwarz Architects.

1 Dickies Arena. One of eight glass-fiber-reinforced concrete panels found on the exterior, integrating themes from the arena's various areas of use; this one celebrating music.

2 Dickies Arena. North entrance façade.

3 Vanderbilt University, West End Tower. Image: Steve Hall, Hall+Merrick+McCaugherty.

4 Vanderbilt University, Zeppos College. Detail of south façade featuring stone-trimmed bay window of the College Great Room and entrance portal at left. Image: Steve Hall, Hall+Merrick+McCaugherty.

Gateway to equal rights

nARCHITECTS

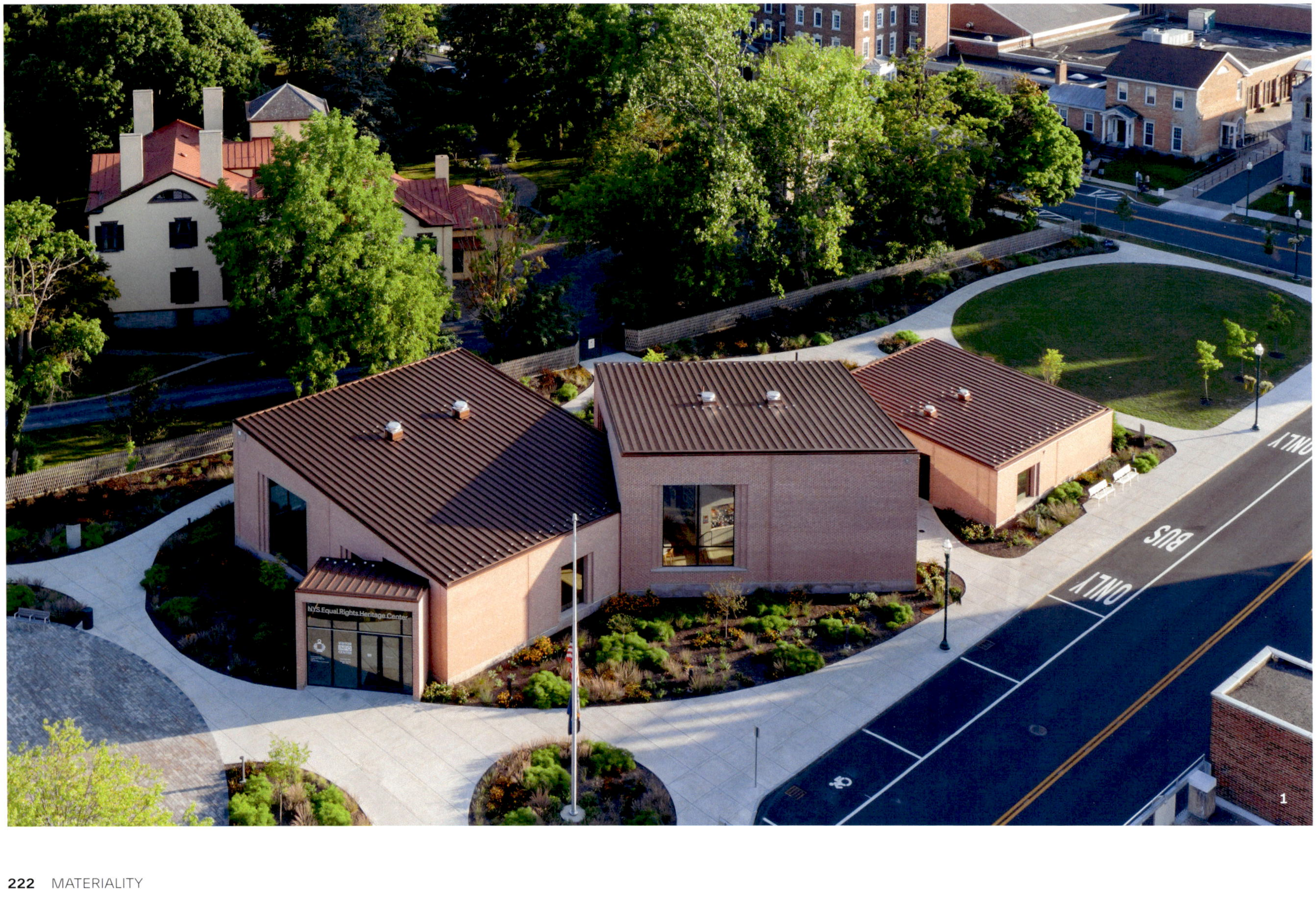

1

In a small but historically significant city in upstate New York, the four angled brick volumes of the NYS Equal Rights Heritage Center hinge open to frame views of the surrounding landscape and adjacent historic sites. Designed by nARCHITECTS and located in the hometown of Underground Railroad conductor and national heroine Harriet Tubman, the Center materializes the intertwining stories of the struggle for equal rights with the use of brick.

Having settled in 1859 after her escape from enslavement, Tubman lived on the outskirts of Auburn for the last 50 years of her life, using the city as a base from which to help other enslaved people find freedom. With the help of abolitionists Frances and William Seward (Abraham Lincoln's Secretary of State and New York State's 12th Governor), Tubman and her husband Nelson Davis built a brick house on South Street, just down the road from the Seward's home, utilizing bricks that were made on-site. Many historically important brick houses from that period and later in the 19th century march along South Street leading towards the project site.

The NYS Equal Rights Heritage Center was designed with attention to this charged historical and material context. Situated at a boundary between the South Street Historic District and a more recent city center disfigured by Urban Renewal in the 1970s, the design of the new Center aimed at reconciling a historical legacy with a more contemporary vision

2

Tubman lived on the outskirts of Auburn for the last 50 years of her life, using the city as a base from which to help other enslaved people find freedom.

for the city's future. Drawing from its historic context, the building's form pays particular attention to proportion with large glazing shaped to echo adjacent rooflines and fine brickwork detailing. As part of nARCHITECTS' design research, brick colors and patterns from the houses along South Street were sampled. Important architectural details were reinterpreted in a contemporary way, including the brick 'shutters'—a progressive thinning of the cast-in-place concrete wall thickness at window jambs, allowing a more expansive and immersive view of the outdoors, while reinterpreting wooden window shutters prevalent in nearby historical houses.

Commissioned by the City of Auburn and the New York State Office of Parks, Recreation and Historic Preservation, nARCHITECTS led a multidisciplinary team to design the center, as well as public outdoor spaces. The new one story, 7,500sf building frames views of important historical landmarks, including the adjacent Seward House Museum —home of William and Frances Seward and a 'stop' on the Underground Railroad— as well as the nearby Memorial City Hall and Westminster Church. In this way the immediate context is integrated with the exhibition content. Celebrating Auburn's and New York State's progressive history of promoting social and equal rights, the exhibition content focuses on abolition, the Underground Railroad, women's suffrage, and LGBTQ rights, with a goal of acquainting visitors with the many attractions connected to equal

3

4

1 Aerial view of NYS Equal Rights Heritage Center. Image: Iwan Baan.

2 Portrait of Harriet Tubman, photo by Harvey Lindsley. Public domain.

3 Portrait of William H. Seward in 1844, by Henry Inman. Public domain.

4 Portrait of Frances Adeline Miller Seward in 1844, by Henry Inman. Public domain.

5 Inside NYS Equal Rights Heritage Center. Image: James Ewing

5

rights throughout the state. One of these elements—the Social Justice Table—brings people together around a circular video display highlighting legislative milestones. The building's four volumes also includes a Taste NY store, public restrooms, and offices for the Auburn Historic & Cultural Sites Commission and local Business Improvement District.

Visitors to the center enjoy the permanent exhibition while connecting to historic Auburn surrounding them, reconciling a history of progress with challenges the country continues to face today.

"The Equal Rights Heritage Center is another example of the importance that New York State places on educating and recognizing its history and honoring those who served as pioneers before us so that we can become trailblazers for future generations," said New York State Office of Parks, Recreation and Historic Preservation Commissioner, Rose Harvey.

The Social Justice Table brings people together around a circular video display.

Brick and the mono-medium ethic
Nader Tehrani

The prospect of working in a single medium is a challenge that some more unconventional practices have taken on throughout history, the most notable in the 20th century being Sigurd Lewerentz and Eladio Dieste. Our own Casa La Roca serves as an homage to their brick research, though advancing it in specific ways. The mono-medium approach defies traditional architectural thought, in part, because of the immediate practical differences between architectural elements—the ground we walk on, the wall that supports a building, the roof that protects from the elements. The performative requirements of these elements have produced a history of material research and development that has assigned optimal materials to each. For this reason, the repellent qualities of ceramic tiles have been oriented toward roofing and wet rooms, while the sturdiness of stone and concrete with foundations, grounds, and hearths. As such, working within a single medium runs against conventional logic and typical methods of construction.

1

Material Intelligence

There are significant exceptions to this rule, and they are often associated with extraordinary constructions where architecture stems more directly from a local resource, be it the stone of Matera, Italy, or the forests of Kiso, Japan. If the former site produced wondrous archaic excavations, where architectural spaces are extracted from rock, the latter produced an elaborate culture of wood details, purged almost entirely of other mediums, creating a syntax of elements that dispense with extraneous mediating joinery. The intelligence of these works stems almost entirely from inventions in wood, with linguistic delight to match. To achieve this, centuries of research went into the extraction of resources, transport methods, the testing of different woods, tooling protocols, and the categorization of joinery details under the stresses of compressive, tensile, and lateral forces. Much work has also gone into finishing processes; differentiating between the interior and exterior, and protecting against the elements and forces of nature.

2

The language, discipline, and ethic of brick

At St. Peter's Church at Klippan in Sweden, Sigurd Lewerentz adopted brick in its many guises, and with it the emergence of a language, a discipline, and an ethic. We recognize that its steel structure serves merely as a framework for the brick enclosure; the floors, walls, vaults, and certain key attributes are formed of brick, while allowing the introduction of glass, wood, and steel for critical interventions (windows, furnishings, handrails). However, Lewerentz's use of brick reminds us of the seminal role of mortar. In his refusal to cut the brick (an ethic he imposes on the medium), he was also subsequently required to develop a malleable bonding strategy: the bonding dimension was sometimes called on to increase beyond common reason, leaving the brick enveloped in a sea of mortar, inverting the traditional figure-ground relationship, an invention that was entirely unique to Lewerentz.

Lewerentz's use of brick reminds us of the seminal role of mortar. The bonding dimension was called in to increase beyond common reason, an invention that was entirely unique to him.

The sum of its parts

The floor of the church is cast in the same bricks as its walls and vaults, and as such biases us toward a monolithic reading of the building as a whole. The space (even if the result of aggregated parts) appears to be excavated out of a stack of bricks, even more so as one inspects the tell-tale details of the ground: the altar and the 'hill' are both construed of the same medium, the former a result of brick rotated in a stacked formation, and the latter the result of the conventional running bond deformed to create a topography that allows for a subtle prospect, a vantage point toward the altar. The associated excavation in front of the 'hill' reveals the bonding system, its depth, and the appearance of the material constitution of the building at large—all brick.

Radicalizing the reach of one medium

In the face of architectural challenges—the navigation of the ground, the compressive forces at work in the walls, and the geometric parameters of the vaults—Lewerentz elects to radicalize the reach of one medium, effectively

3

By essentially eliminating the need for other trades, matters of cost, coordination, and craft are redirected toward one mission—brick assemblies.

demonstrating where architectural discipline possibly lies: somewhere in the confines of constraints, and less so in the limitless categories of difference. Here, he calls on those differences to be located within operations onto one medium, erasing the evidence of elemental categories to the eye. Instead, he reveals a nuanced shift in material behavior through the sites of its interaction with the human body: the ground onto which we step, the furnishings we touch, and the tactile interface that rituals require.

Most importantly, Lewerentz identified the way in which labor is harnessed in its most strategic capacity to create a cerebral opulence, where the craft is allowed to remain raw and crude. By essentially eliminating the need for other trades, matters of cost, coordination, and craft are redirected toward one mission—brick assemblies—and with it the rewards are not only of the generosity of its affect, but of the advancement of a discipline.

Mediating the medium

Still, the insistent address to a single material medium is fraught with added complexities, most of them reconciled through architectural amendments and predictable

mediations. In Matera, where stone cannot offer human comfort—furnishings, fabrics, and finishes come to the rescue; meanwhile, black terracotta tile (Kawara) protects the Japanese temple's roof from the elements. It is all the more poignant, then, how Lewerentz was able to display an intellectual insistence that went beyond the positivism of solving problems with brick as his medium.

Ethic of purity

The predisposition toward the singularity of a material medium may ultimately be an artifice. One could claim it is a strategic device, in effect, to build adversity from within one's approach, if only to demonstrate the ingenuity of architecture in response. And in the appeal to parasitic details in support of the host material, the ethic of this philosophy remains intact, precisely because the excursions outside of the medium make no claims to dominance. Behind the mono-material sensibility is an assumed ethic of purity, and the difficulties that underlie realizing it are taken up as badges of honor when overcome. Yet, it is not uncommon that the assumed simplicity of single-minded decisions leads to complications. Thus, despite my own conceptual underpinnings (or maybe precisely because of them) the idea of hybridity has become an innate part of the tectonic riddles that I confront. With Casa La Roca being a project we have redrawn for 25 years, we have also expanded our research through other explorations in the Tongxian Gatehouse, the Rock Creek House, and RISD's North Hall—all of which push the limits of this unique medium.

1 Casa la Roca, Office dA.

2 St. Peter's Church, Björkhagen, Stockholm, Sigurd Lewerentz. Image: seier+seier.

3 Tongxian Gatehouse, Office dA. Image: Dan Bibb.

4 RISD North Hall, NADAAA. Image: John Horner

5 Rock Creek House, NADAAA. Image: John Horner.

Innovations in computational design for curved brick construction

HDR

For thousands of years, brick has been a popular choice for construction due to its durability, strength, and versatility. With the development of contemporary detailing techniques and digital modeling, brick buildings have become even more aesthetically appealing and functional. This technical paper considers the latest trends, best practice, benefits, and the challenge of incorporating contemporary detailing techniques into brick construction projects. Specifically, it examines their application in the creation of the Brendan Iribe Center's Michael Antonov Auditorium at the University of Maryland.

Contemporary detailing techniques have revolutionized the world of brick construction, unleashing a realm of captivating patterns, textures, and colors that empower architects and builders to create truly remarkable designs. These techniques merge the realms of aesthetics and functionality, enabling the realization of brick structures that excel in energy efficiency, sound isolation, weather resistance, and sustainability. By harnessing the potential of contemporary detailing, architects and builders can unlock a world of endless possibilities, where form and function seamlessly coexist and bricks transcend their traditional role to become works of art.

Expanding the potential of brick

The future of brick construction is being shaped by these innovative techniques, paving the way for extraordinary architectural achievements that captivate the eye, enhance comfort, and honor the principles of sustainability.

Patterns: Brick bonding patterns are the foundation of this creative exploration. Techniques such as the timeless Flemish bond, the classic English bond, or the modern stack bond enable architects and builders to achieve varying visual effects while maintaining structural integrity. Designers can also leverage different brick sizes and shapes to introduce complexity and intricacy to their designs, fashioning mesmerizing patterns that capture the eye and spark curiosity.

1

Contemporary detailing techniques have revolutionized the world of brick construction, unleashing a realm of captivating patterns, textures, and colors.

Texture: Texture plays a crucial role in adding depth and tactile appeal to brick surfaces. Manipulating brick texture with contemporary detailing techniques offers opportunities to introduce roughness, smoothness, or even intricate relief patterns.

Colors: In contrast to the traditional red or brown bricks, modern techniques enable the use of an expansive palette, transforming the aesthetic possibilities of brick construction. This newfound versatility is achieved through various methods, such as incorporating pigmented or glazed bricks and the application of different coatings to the surface of the brick. Pigmented bricks open a world of color options, creating striking monochromatic or multicolored façades.

Additional innovations in energy efficiency and thermal performance, sound insulation and noise transmission, and weather resistance further expand the instances where brick can be used.

2

Using algorithms to develop brick patterns opens new avenues for creativity.

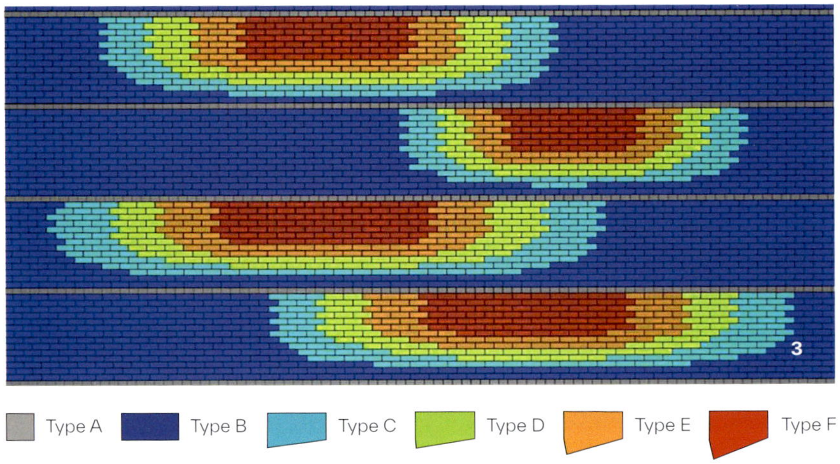

3

Combining computational design and contemporary detailing

Contemporary detailing involves meticulous attention to the small-scale aspects of a building, such as joints, connections, materials, textures, and finishes, intending to achieve specific functional, aesthetic, and performance goals.

Contemporary detailing techniques for brick have made a great leap forward with the integration of algorithms into the design process. This offers architects and designers a powerful tool to create efficient and visually captivating brick patterns. By harnessing the capabilities of computational design tools, the possibilities expand for creating intricate designs that are both visually-striking and cost-effective to construct.

Using algorithms to develop brick patterns opens new avenues for creativity and efficiency. Architects and designers can define various parameters, such as the available brick sizes, the desired aesthetic effect, and the structural considerations of the building. Using computational algorithms, these parameters can generate patterns that maximize efficiency in terms of material usage, construction time, and structural integrity.

Integrating algorithms in brick pattern design brings numerous benefits to the construction industry. Architects can explore a vast array of design options, experiment with intricate and complex patterns, and visualize the result with greater accuracy and iterations. Furthermore, the optimization capabilities of algorithms ensure that the resulting brick patterns meet structural requirements while minimizing waste and reducing construction costs.

One example of this is the use of parametric design tools to create complex brick façades. By inputting the desired façade dimensions, brick size, max. expansion joint spacing, and other parameters, the algorithm can generate a brick pattern that is efficient to construct, minimizing the amount of cutting or custom brick pieces required. This can save time and money during the construction process, while also reducing waste.

The use of algorithms in developing efficient brick patterns is still a relatively new development, and there are challenges associated with this approach. For example, the complexity of the algorithms can make it difficult to communicate the design intent to builders and contractors who may be less familiar with computational design tools. Additionally, the use of custom brick pieces can add to the cost of construction. Despite these challenges, the use of algorithms in developing efficient brick patterns holds great promise for the future of brick construction. By creating optimized patterns that are both efficient and visually appealing, architects and designers can push the boundaries of what is possible with this timeless building material.

1 Brendan Iribe Center's Michael Antonov Auditorium at the University of Maryland. Image: © Dan Schwalm.

2 3D render showing parametric brick wall design development.

3 Michael Antinov Auditorium—parametric brick wall design development.

4 Interior of auditorium, showing the parametric brick wall. Image: © Dan Schwalm.

5 Detail of parametric brick wall. Image: © John T. Consoli.

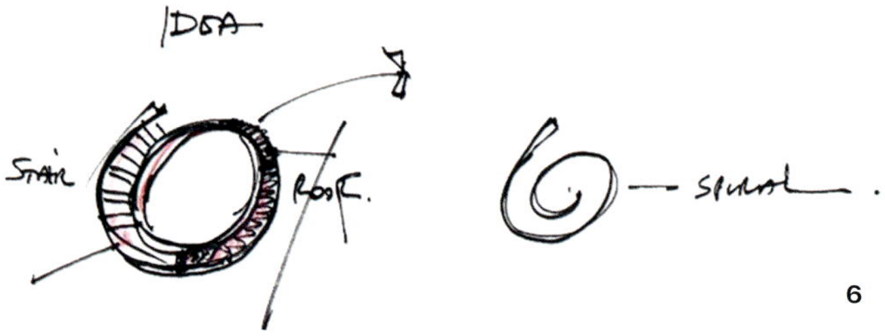

6

Approaches to creating curved brick walls

Brick in curved and sloping walls can provide a high degree of structural stability. The weight and density of the bricks can help to resist lateral forces, making brick walls a strong and durable option for curved and sloping walls. Additionally, the smooth, flowing curves that create a sense of motion and fluidity create aesthetic appeal, while the use of different brick patterns and colors can create a unique and striking visual effect.

So how is it done? One approach to creating curved walls with sloping profiles is to use a technique called 'brick corbeling'. This involves gradually overlapping bricks as they are laid, creating a stepped profile that follows the curve of the wall. By carefully selecting the size and shape of the bricks and the degree of overlap, builders can create a smooth, continuous curve that also slopes.

Another approach is to use specially designed curved brick units, which are prefabricated to a specific radius and angle of slope. These units can be used to create walls with a uniform curve and slope, without the need for corbeling. Curved brick units can also be combined with standard bricks to create intricate patterns and textures.

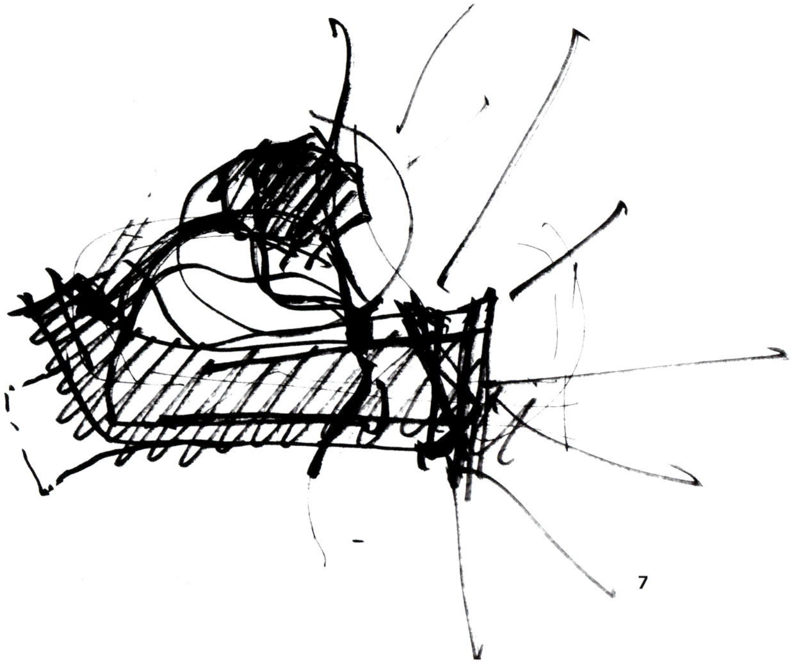

7

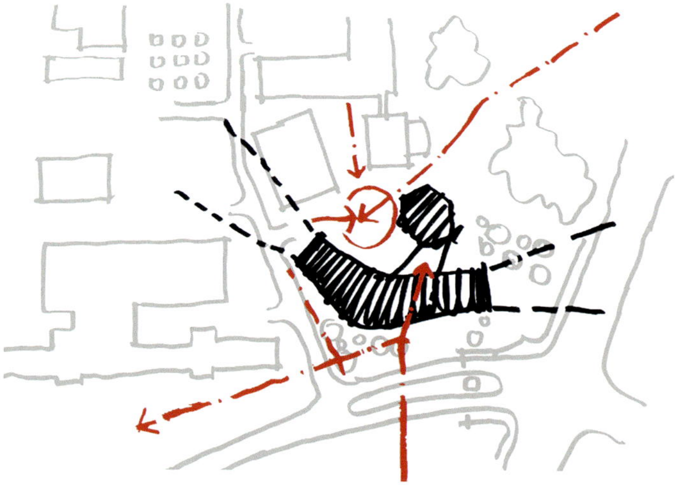

8

6 Audit concept sketch for the Michael Antonov Auditorium.

7 Michael Antonov Auditorium, conceptual plan.

8 The University of Maryland, Brendan Iribe Center, site plan diagram.

Challenges in designing curved brick walls

Regardless of the technique used, constructing curved and sloping brick walls requires a high level of skill and attention to detail. Careful planning and detailed precision and craftsmanship are essential to ensure that the curvature and slope of the wall are consistent and accurate and that the bricks are laid with the correct degree of overlap to maintain the structural integrity of the wall.

Expansion control is an important consideration. Due to the curvature of the wall, there may be differential movement between the brick and the substrate which can lead to cracking or other forms of damage if not properly addressed.

One way to control expansion is to incorporate expansion joints. Expansion joints are gaps in the brickwork that allow for movement and expansion without damaging the brick or the substrate. These joints can be placed at regular intervals along the length of the wall and should be designed to accommodate the expected movement of the wall. Another way to control expansion is to use flexible materials in the construction of the wall. For example, using a flexible mortar that can accommodate movement can help to reduce stress on the brick and the substrate. Similarly, using flexible flashing, connections, and weep holes can help to prevent water infiltration and damage to the wall.

Temperature changes on the wall should also be considered. Brick is a material that expands and contracts with changes in temperature, so it is important to take this into account when designing and constructing curved walls. Thermal expansion joints, which are similar to expansion joints but are designed to accommodate changes in temperature, can be used to control the effects of thermal expansion on the wall.

In addition to these measures, it is important to ensure that the wall is properly designed and constructed to minimize stress on the brick and the substrate. This may include using appropriate reinforcement, ensuring adequate drainage and ventilation, and ensuring that the wall is properly supported and stabilized.

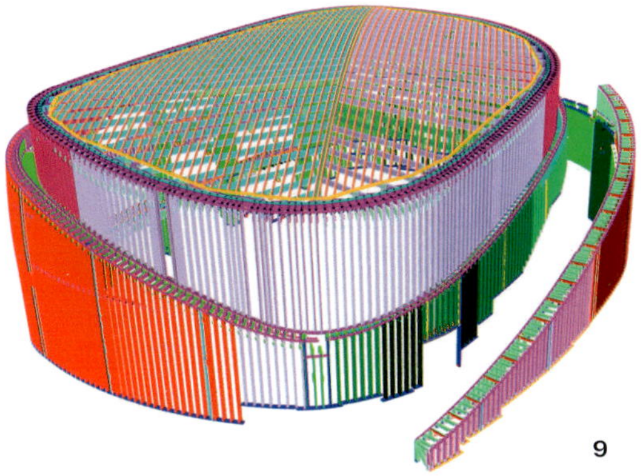

9

9 CAD image showing the structure and cladding of the Brendan Iribe Center's Michael Antonov Auditorium. Image: courtesy of radiustrack.

10 Brendan Iribe Center's Michael Antonov Auditorium.

Auditorium at the Brendan Iribe Center

When working on the brick detailing at the University of Maryland's Brendan Iribe Center, we utilized two different parametric detailing techniques to define and detail the form. The first technique involved incorporating the expansion joints and rhythmic pattern for curvature in both the plan and section. The second technique focused on module patterning, allowing for the acceptance of the curve and bay expansion location in the plan while serving as an art element within the curved elevation.

At the Iribe Center, the design team employed advanced algorithmic software to meticulously design the pattern and spacing of the brickwork, strategically incorporating 22 expansion joints to accommodate the curvature and brick distribution. The algorithmic approach enabled precise determination of the vertical joint locations and spacing, taking into account the number of bricks and the overall curvature of the structure. The design process entailed working with 22 distinct curves and a singular convex shape, meticulously identifying specific locations, and organizing patterns to facilitate accurate brick allocations for the contractor. This algorithm-driven detailing approach ensured an exact and well-coordinated execution of the brickwork, achieving the desired visual and structural outcomes for the project.

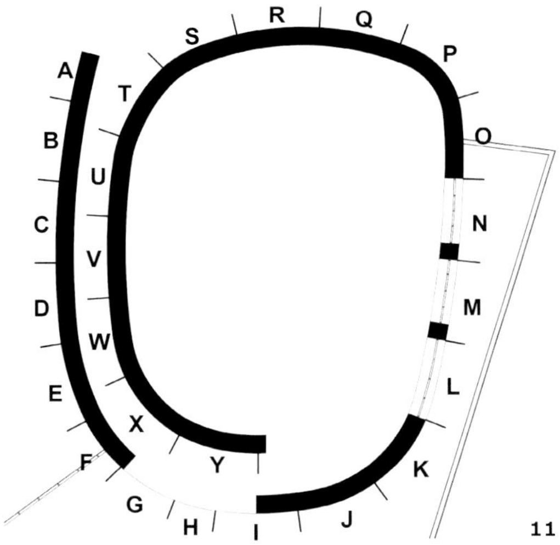

11

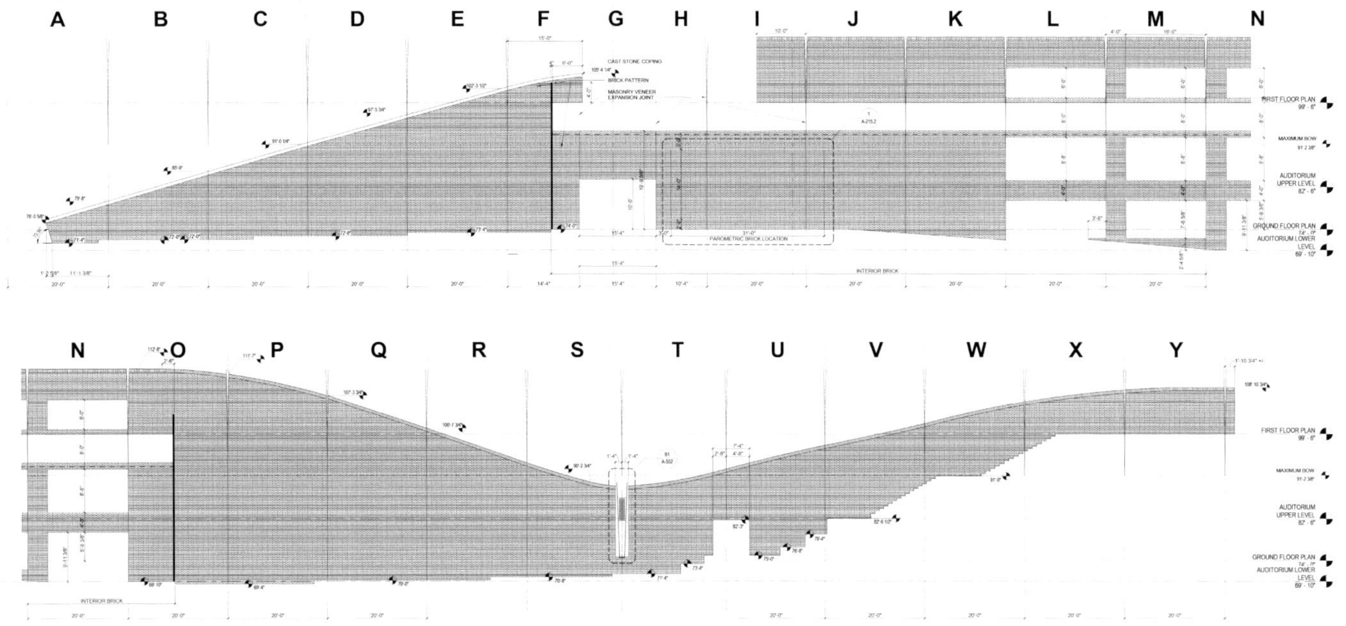

12

11 Auditorium brick diagram, Brendan Iribe Center.

12 Auditorium flattened elevation, Brendan Iribe Center.

13 Brendan Iribe Center for Computer Science and Engineering. Image: © Dan Schwalm.

13

I love bricks
Morris Adjmi

1

As an architect, I've chosen brick for many projects and used them in many different ways. If I had to limit my material palette to just one, I'd choose brick. Perhaps it's unfair to consider brick as a single material because it has such a range of possibilities—shape, size, color, pattern, surface, finish, etc. In addition to the cornucopia of options, I also see brick as a connection to the past and a harbinger of the future (constructions).

Material

Bricks have been in use since 7000 BC and were first made of mud that was slowly dried in the sun. The connection to the earth and human's ability to create them is significant. Bricks connect us to nature but also to each other through their recognizable form and scale. For many centuries, bricks have been primarily made of fired clay, although increasingly we are seeing new materials such as reclaimed construction waste and recycled glass and plastic that go some way toward addressing the climate crisis. An early example of this approach is the Heineken 'WOBO' (World Bottle) brick—a custom piece of emerald green glassware made to serve as both a beer bottle and a building material. MA explored the idea of a sustainable brick on Wilf Hall, a LEED Platinum project for New York University that features a façade composed of bricks made with industrial waste previously destined for a landfill.

2

3

4

1 View of the WOBO building, 1964. Image: Courtesy of Heineken Collection Foundation, image # 11504, ©Heineken Collection Foundation.

2 Collection of Heineken World Bottles. Image: Courtesy of Heineken Collection Foundation, image # wobo08, ©Heineken Collection Foundation.

3 World Bottle. Freddy Heineken asked the architect John Habraken to draw up these stackable bottles in 1962. Image: NAI Collection, VOOV 195, Wikicommons.

4 NYU Wilf Hall by MA | Morris Adjmi Architects. Image: Matthew Williams.

Shape

We see bricks as part of a whole. The individual units are usually rectilinear and laid up in different patterns as a result of a desire for stability or functionality. The modular shape of bricks allows for a variety of patterns but also fast and simple construction. We also see a variety of special shapes that are used to frame windows and doors or create cornices that crown buildings or ribbon courses that separate floors. It is the combination of the regular and special shapes that add consistency and welcome decoration to building façades. MA recently completed the Grand Mulberry, a building that uses custom handmade bricks featuring domes to create a spectral reference to the traditional tenement buildings found throughout New York City's Little Italy neighborhood. Like pixels on a screen, we see how bricks have an unlimited potential to delight!

5 Detail of bricks used for the Grand Mulberry by Morris Adjmi. Image: courtesy of Glen-Gery.

6 Detail of bricks used for the Grand Mulberry by Morris Adjmi. Image: courtesy of Glen-Gery.

7 Santa Maria del Fiore Dome. Image: Adobe Stock.

Size

A building as a block in our hands. The endless possibilities that many of us discovered playing with LEGO® can be found in brick construction. The principles of the size of the unit combined with the hand-applied scale make the brick a convenient and powerful material. It gets even more exciting when we explore the possibilities of different sizes like modular, Roman, and Norman. The power of bricks comes from their individuality and their common purpose: E pluribus unum!

Pattern

Opus spicatum, meaning spiked work, is the Roman term for bricks or stone set in a herringbone pattern. This pattern can be found on walls and floors and shows the effect that simple patterns can have on the visual effect of rotating or realigning the direction of bricks. It is interesting to note that this pattern was used by Filippo Brunelleschi in covering the previously un-solvable problem of constructing the dome of Santa Maria del Fiore in Florence. This example shows that beauty and function come together elegantly. Patterns are endless and enhance and contrast with the simplicity of the brick unit.

Bricks typically reflect the clay available in the location where they are manufactured.

9

8

Color

When we think of bricks, we naturally think red, yet bricks typically reflect the clay available in the location where they are manufactured. Used to construct buildings, they help to define neighborhoods and even entire cities. Consistency of color can create a backdrop, a mood, and a sense of place. Color can also reflect a particular time and place. The Ladies Mile Historic District in New York City, roughly 14th to 23rd Streets and Broadway to 6th Avenue, was primarily rendered in white (both brick and stone). This was a result of those New York and Chicago architects who were active during the Chicago World's Fair of 1892—a story that was documented in *The Devil in the White City*. All of the buildings in the Fair were intended to be faux painted and multicolored to simulate real buildings but in the end were all painted white due to time constraints. As a result of the impression of the 'modern city' many of the architects that were active at the time began to use light bricks which dramatically affected the neighborhoods and cities where they were working.

Surface

Bricks can be handmade, extruded, wire cut, or smooth. The resulting surfaces, coupled with sheen—glazed, matte, or natural—add to the overall effect of the individual brick but more so the building façade. The options are further enhanced by the way that these surfaces are combined with the other characteristics mentioned above (shape, color, size, etc.) to produce an endless number of options that allow architects to design buildings that establish a connection to our shared history and create a sense of place.

Mortar

Mortar binds bricks together but also provides further ways to distinguish the surface. Through color—contrasting or matching—mortar creates a consistency of surface. The color of mortar also has the ability to accentuate the individual bricks or the overall surface. I like to think of the size and shape of the joints as the eyeliner setting the tone for the party. Just look at the refined penny joints of federal town houses in New York compared to the big and loosely applied mortar used by Lewerentz on the St. Mark's Church in Björkhagen, Stockholm. As we look at brick surfaces we understand how we build, and even more about how we live.

11

10

8 The Sterling Mason by MA | Morris Adjmi Architects. Image: Field Condition.

9 Martin Creed, Work No 1812. Image: Google Arts and Culture.

10 Interior of Markuskyrkan (St. Mark's church), Björkhagen, Stockholm. Built 1958-60. Architct: Sigurd Lewerentz. Image: Arild Vågen, Wikicommons.

11 Detail of brick and mortar, Markuskyrkan i Björkhagen, Stockholm. Image: Holger Ellgaard, Wikicommons.

12

Bricks have proven their uncanny ability to present themselves in the concrete world and reflect metaphorical aspects of our culture.

Bricks have also proven their uncanny ability to present themselves in the concrete world and reflect the metaphorical aspects of our culture. They delight in their simplicity and ease of construction but also have unlimited abilities to mirror the milieu of the time and place. An unexpected experience with brick that has stayed with me was visiting Pablo Gómez Uribe's exploration of bricks as a central element of his artistic vocabulary in the 2018 exhibition titled *All that is Solid* at PROXYCO gallery in New York City. The exhibition's initial piece, 'Tridimensional Palimpsest', comprises 200 bricks crafted from pumice stone, intricately tracing the architectural memory of PROXYCO's former floor plan.

Morris Adjmi, Fellow of the American Institute of Architects (FAIA) is Principal & Founder of MA | Morris Adjmi Architects.

12 Exterior view of Markuskyrkan i Björkhagen, Stockholm. Image: Holger Ellgaard, Wikicommons.

13 Detail from Pablo Gómez Uribe's exploration of bricks, *All that is Solid,* PROXYCO gallery, New York City. Images: courtesy of PROXYCO.

14 *All that is Solid,* PROXYCO gallery, New York City. Images: courtesy of PROXYCO.

Preserving and restoring brick

Don Foster

1

Regardless of their size, brick buildings are usually considered strong, safe, stable, and durable. Sturdy structures such as these, however, are in constant motion. For example, as external temperatures change, brick expands when heated and contracts when cooled. Beyond just brick, mortar and many other materials make up the wall system, all of them subject to movement. As strong as these buildings are, they still need to be maintained. The preservation and restoration of the external brick component is important as these buildings age over time.

Repointing brickwork

Brick and mortar are the two main products where brick is used to clad a building. In most cases, brick will outlast the mortar joint itself, so a crucial part of preserving the building is repointing the mortar as needed. Since it is rare that an entire building will require repointing, only designated areas of wear to the mortar need to be addressed. It is simpler to repoint mortar than it is to replace broken brick. Today, repointing is made easier by testing for the original mortar used in order to achieve the right mix.

2

Cracking and spalling

Damaged brick is a bigger issue and more difficult to resolve depending on the cause of the damage or the problem it presents.

Cracked brick running down a wall indicates the building may have settled. An engineer's involvement is necessary to diagnose the problem and embark on a detailed restoration project. This is vital in the case of spalling where the surface shears off, or an area of a building's brick surface disintegrates. Establishing the cause of the spalling is essential, and a solution must be devised and carried out.

For small areas of spalling, investigation may often reveal that the underlying problem is worse than that of a few spalled bricks. Typically, quite a few bricks in the same area will be compromised. It's important to note that remedial builders might not be able to estimate the extent of a problem until they open up the wall area that needs remediation.

Remediation

Today, our industry has the best architects, engineers, and restoration and masonry contractors available, and they do some remarkable work restoring aging buildings. Indeed, their work is key to the masonry industry's reputation and its claim on the best building product.

Surprisingly, the hardest challenge is not the construction phase, but the visible aesthetics when a restoration job is complete. This is sometimes difficult to achieve because materials necessary for a good-looking repair or match are limited by what is available. Several brick plants that produced millions of brick in earlier years are now no longer in operation. Because of this, it is sometimes

What makes the stain permanent is that it absorbs into the pore structure of masonry.

difficult or impossible to locate an identical brick to use for repairs. Sometimes a brick stain may be required—one that withstands the test of time, and which neither harms nor compromises brickwork. Selecting the right stain is vital.

Brick staining

Color matching Matching a historical brick and also the mortar can be extremely difficult. Some common sense rules apply. First, if the brick is no longer available, find the closest brick in size and texture. Next, get as close as you can to the base color of the brick. Lastly, mortar plays just as important a role as the brick. If the brick is modular in size (2¼ x 7⅝ inches) mortar can represent as much as 18 to 20 percent of the wall area. This will, in turn, have a major effect on the color match. Matching the mortar color, texture, and the type of joint that is on your building will make all the difference in quality and appearance.

The stain product selected for either the brick or mortar is a critical part of long-term performance, and knowledge of the different stain products in the marketplace today will be key to the success of a durable solution.

Clearly, it's important to be aware of inaccurate information. Some products market themselves as a proprietary brick stain, when in reality they may consist only of a watered-down paint product. While the product may tout great technical data, lab tests, and support information, it may not have long-lasting field performance to support such claims.

In times past lab testing was important, but this is no longer the case. Without historical data on projects, there is a dearth of history to represent claims of a product's field performance. Instead, a range of indicative tests are cited to support claims to product reliability in the long term. Going forward, the potential damage to brick buildings through the use of improper staining products, especially to beautiful historic buildings, is something to avoid at all costs. Staining applied correctly with the right product can be successful and greatly rewarding to all involved.

Time in the industry Some stain companies have been in business for up to 35 years, with a few claiming a much longer history. A number of these may have been painting companies who, in times past, would have painted the brick on a house as opposed to calling it a brick stain. Companies with a history of 30 years or more should be able to provide warranties and refer you to hundreds of successful jobs that demonstrate the performance of their products over time. Caution is advised when considering a newer company that has been around only a few years, yet is offering a 50-year warranty. Are there proven tests that could support that claim?

Brick paint or brick stain? The difference between brick paint and brick stain, from this writer's point of view, is simple. A brick stain should be a translucent-in-nature slurry applied by brush to the absorptive brick, one brick at a time. This stain embeds itself into the pore structure of the brick and, most importantly, allows the straight-from-the manufacturer maintenance-free brick to remain a maintenance-free brick. It can get soaked by the rain, yet dry and 'breathe' naturally after it has been stained, just like an unstained brick.

On the other hand, a brick 'paint-like' stain contains a percentage (any percent) of latex or acrylic, consists of either paint or watered-down paint, and is re-labeled and called a 'proprietary brick stain'. This product, once applied, turns a maintenance-free brick into a maintenance one. These watered down paints can be sprayed, rolled, or brushed on, and sometimes all three applications could be used on the same project. Unfortunately, we have seen these 'stains' either fade or turn into a chalky whitish film that can compromise the brick, its appearance, and its maintenance for years.

Case studies in staining and modifying color

Mishawaka City Hall The old Liberty Mutual Insurance building was built in the 1980s. In 2021, the building was renovated for the new 93,500 sq ft Mishawaka City Hall building. For the $20-million renovation, 8,000 new bricks were added for infill and replacement bricks. The brick added was the Belden Commodore Full Range brick, size closure, 3⅝ x 7⅝ x 3⅝ inches.

The old and new bricks were stained to a coffee color for a cohesive look. An issue we ran into was that a sealer had previously been applied. What makes the stain permanent is that it absorbs into the pore structure of masonry, so sealed masonry prevents a successful Masonry Cosmetics stain. The sealer had to be removed, and even then we were concerned that the outcome on a large building such as this would be problematic.

1 The University of Notre Dame's 'Golden Dome" Administration Building, 2015. The preservation and restoration of this building contribute to its longevity and recognition as a heritage building and landmark.

2 University of Notre Dame Power Plant, showing both the original building and the addition with the unifying effects of brick staining.

3 Mishawaka City Hall before addition and brick staining, 2021.

4 Mishawaka City Hall after addition and brick staining, 2022.
All photos: Masonry Cosmetics, Inc.

One significant challenge was damage to the brick color caused by deeply embedded atmospheric dirt.

5

6

7

University of Notre Dame Main Building The notable Golden Dome Building was constructed in 1897 and serves as the administrative center and offices for the university's president and provost. The yellow clay bricks are known as 'Notre Dame bricks'.

Restoration on the building commenced in 2015, however, the original brick wasn't available and was hard to match. The university had some bricks from a previous deconstruction and while it was a close match in size and texture, the color of the brick differed. Masonry Cosmetics was commissioned to stain the brick, mortar and limestone bands. One significant challenge was damage to the brick color caused by deeply embedded atmospheric dirt. The contractor's application process required a clean, dry surface so that problem areas were cleaned multiple times. Great care was taken not to damage darkened brick by overcleaning.

5 Column on the University of Notre Dame main building with damaged bricks that need to be replaced.
6 Column before brick replacement.
7 Column after brick staining.
8 Brick work before restoration.
9 Brick work after restoration.
All photos: Masonry Cosmetics, Inc.

10

The original
structure was made up of five different runs of brick.

University of Notre Dame Power Plant In 2008, a new three story addition was added to the University of Notre Dame's power plant, as part of the university's sustainable energy plan to reduce their carbon footprint. The addition was made up of 80 percent Spring Hue Velour texture module and 20 percent Concord Velour. The original structure was made up of five different runs of brick.

By creating multiple recipes, Masonry Cosmetics stained the five runs of brick and the mortar to match the new addition. The largest struggle was that the existing bricks had multiple different textures—our stain preserves the natural texture and tone variations of the original masonry. To make bricks with different textures and absorbency look the same often requires them to have slightly different recipes. This takes a lot of time, finesse, attention to detail, and organization.

Why stain bricks

Staining has been used to change the color of brickwork in a variety of circumstances, including, but not limited to, the following:

- Correcting brickwork in which multiple brick colors may not have been properly blended, resulting in a patchwork appearance.
- Matching brickwork constructed of brick from different production runs.
- Matching brickwork installed in a building addition. Not only can the new brickwork be matched to the existing brickwork, but the existing brickwork can be matched to the new brickwork.
- Covering surface imperfections of brick due to the presence of chips, lime pops or pyrite pops.
- Repairing brickwork that was damaged due to improper cleaning.
- Updating existing brickwork to change aesthetics yet maintain natural brick appearance.
- Providing an exact color for the desired brickwork when a custom color or color match cannot be achieved through the manufacturing process.
- Modifying both the brick units and mortar joints to achieve a consistent color throughout the facade, to resemble painted brickwork.
- Modifying only the mortar joints to match the color of surrounding mortar joints.

This article by Don Foster, President & Director of Masonry Cosmetics Inc., was sourced from Brick Brief, *The Brick Industry Association, June 2021. For further information on preserving brickwork, see the article in full.*

10 University of Notre Dame Power Plant, before addition and brick staining.

11 University of Notre Dame Power Plant, after addition and brick staining. Photos: Masonry Cosmetics, Inc.

University of Notre Dame—a study in brick

1

Known simply as Notre Dame, the University of Notre Dame du Lac is a private Catholic research university in Notre Dame, Indiana, north of the city of South Bend. It was founded in 1842 by French priest Edward Sorin. Originally for men, the university started accepting undergraduate women in 1972.

In its suburban settings, the University Notre Dame is home to notable landmarks such as the Golden Dome, the Word of Life mural, the Grotto, Notre Dame Stadium (and its famous mural referred to as 'Touch Down Jesus'), and the Basilica. Notre Dame is consistently ranked and admired as one of the most beautiful university campuses in the United States and around the world.

Nineteenth century

The college awarded its first degrees in 1849 and quickly developed a strong reputation for its academic offerings and building program to accommodate the growing student and faculty population. Successive administrations under Patrick Dillon, William Corby and Auguste Lemonnier built on Father Sorin's work. The presidency of Thomas E. Walsh (1881-1893) focused on improving Notre Dame's scholastic reputation and standards.

Twentieth century

In the twentieth century Notre Dame continued to expand, however it was under the leadership of John W. Cavanaugh (1905-1919) that the movement towards a research university gained impetus. Cavanaugh modernized educational standards, boosted student numbers and attracted many eminent scholars, established a chair in journalism, and introduced courses in chemical engineering. During his time as president, Notre Dame rapidly also became a significant

2

force on the football field. Indeed, during the early decades of the twentieth century, one of the main driving forces in the university's growth was its football team, the Notre Dame Fighting Irish. Notre Dame's success mirrored the rising status of Irish Americans and Catholics in the 1920s. Catholics rallied around the team and followed the games on the radio, especially when it defeated teams from schools that symbolized the Protestant establishment in America—Harvard, Yale, Princeton, and Army.

In 1917, Notre Dame awarded its first degree to a woman, and its first bachelor's degree in 1922, although it was not until 1972 that attendance of female undergraduates at the university won general acceptance.

The rise of Adolf Hitler and other dictators in the 1930s forced many Catholic intellectuals to flee Europe; President John O'Hara (1934-1939) brought many of them to Notre Dame where they were welcomed. In their research and teaching the exiles developed a distinctive emphasis on the evils of totalitarianism.

3

Consistently ranked and admired as one of the most beautiful university campuses in the United States and around the world.

Post Second World War

Under the presidency of Theodore Hesburgh (1952–1987), Notre Dame entered a period of massive growth, transforming itself from a school mostly known for its football to a top-tier university, academic powerhouse, and America's leading Catholic university.

In the eighteen years Edward Malloy (1987-2005) was president, the school's reputation, faculty, and resources grew rapidly. John I. Jenkins took over from Malloy in 2005 and vigorously pursued his goal of making the university a leader in research that recognizes ethics and builds the connection between faith and studies. During his tenure, Notre Dame has increased its endowment, enlarged its student body, and undergone many construction projects on campus.

In 2014, John Simpson the eminent London-based architect and a leading practitioner of New Classicism and New Urbanism, was commissioned to design a new School of Architecture. Most fittingly, Smith's work in the United Kingdom has found new avenues to explore and expand into through the commission.

4

5

1 Walsh Family Hall.
2 Alumni Hall.
3 Alumni Hall.
4 Allumni Hall detail.
5 The Basilica of the Sacred Heart.
6 Sorin Hall.
7 Crowley Hall.

The buildings of Notre Dame—past and present—are striking for their use of brick.

Basilica of the Sacred Heart

The Basilica of the Sacred Heart is a Catholic church on the campus of the University of Notre Dame. It was consecrated on August 15, 1888. Built in the neo-gothic style this National Register building (1978) houses the largest collection of 19th century French stained glass in the world, as well as the oldest carillon in North America. The church has 44 large stained glass windows and murals completed over a 17-year period by the Vatican painter Luigi Gregori. The basilica bell tower is 230 feet (70 m) high, making it the tallest university chapel in America.

8 Golden Dome and main building.

9 The Old College.

10 Purcell Pavilion at the Edmund P. Joyce Center.

11 Stayer Center for Executive Education (2013). Home of the Mendoza College of Business Executive Education department which supports and serves leaders and other working professionals through the Executive MBA, open enrollment and custom programs.

12 The Notre Dame Biolchini Hall of Law, established in 1869, is the oldest Catholic law school in the country and one of the oldest university law schools in continuous operation. The building was opened in 1930 and has received two additions.

8

9

The Golden Dome

The Main Administration Building (or the 'Golden Dome') hosts the administrative offices of the university, as well as classrooms, art collections, and exhibition spaces, and is listed on the National Register of Historic Places.

The Golden Dome is immediately recognizable and creates a striking effect. Three buildings were built at the site—the first in 1843 which was replaced with a larger one in 1865 (destroyed by fire in 1879), followed by the third and current building completed in 1882.

The Old College

Old College was built in 1843 by the founder of the University of Notre Dame, Rev. Edward Sorin, C.S.C., and the rest of the Holy Cross brothers. It is the oldest standing building on campus and like other historic structures of the university is a listed building.

10

11

12

The development of the university campus through its carefully considered masterplan has remained faithful to the conviction that a building remains worthy of its workers.

3 The blended stockpiles can reach up to about 8 metres high to allow the Loader to cut through the face of the stockpile, while the individual clay/shale stockpiles can a be higher.

4 Uncrushed shale-clay.

5 Layered dirt in quarry.

Stage 2: Primary crushing

Unprocessed dirt from a selected outdoor stockpile is scooped up by a front-end loader and carried to a large bin. At this stage, the dirt is a mix of small fragments, as well as clay and lumps of shale weighing up to 30 kg or more. Primary crushing breaks the raw material down into smaller particles ranging from 40 to 180 mm in size.

The material drops into a powerful CVR primary crusher which can pulverize about 250 tonnes of material per hour.

6 Brickworks' newest front-end loader is a Caterpillar 990 K and the largest owned by the company. Its bucket can hold up to 24 tonnes of clay. As compared to Brickworks' standard loader (988 K/H) with a bucket with a clay-holding capacity of 12 tonnes.

7 The recently installed feeder bin at Plant 2 has a 100-tonne capacity, which explains the need for a front-end loader with a substantial loading capacity.

8 From the primary crusher, a conveyor (the first in an elaborate system of conveyors that service different stages of the brickmaking process) carries the dirt to the undercover storage area. This area contains clay storage bays.

Stage 3: Souring the clays

In each of the storage bays, an overhead conveyor moves horizontally, backwards and forwards, depositing a steady stream of clay in layers. This inside stockpile builds to a height of about six meters and six meters in width. Protected from the elements, the clay is allowed to further sour.

The strategic stockpiling of raw materials in sufficient quantities means that brick production is able to continue without interruption. The entire storage area has enough material to maintain production for many days of plant operation if stoppages occur at the primary crushing stage.

After a few days the loader scoops up fully-soured dirt from the bay. The machine operator positions the bucket at the bottom layer and in one single, upward sweep fills the bucket. This upward sweeping action ensures that the blend which is about to enter the grinding process contains a mix of layered clay and shale types.

9 Plant 2 has nine storage bays, each with a capacity of approximately 7,000 tonnes.

10 Front end loader scooping dirt.

11 The new 100 tonne bin that supplies clay and shale to the sieving and mixing processes

12 On the conveyor heading into the Incla pans.

13 Incla pans - also called pan mills or dry pans.

14 Exiting the Incla pans.

12

Stage 4: Grinding and pre-mixing the clays

Grinding The loader tips the dirt into a feeder bin, which drops the material onto a conveyor below. The new feeder bin is capable of holding around 100 tonnes of clay.

From there it travels to a hammer mill or secondary crusher where it undergoes another round of crushing.

Oversized clumps are diverted to a holding bin which progressively discharges the larger fragments of shale into circular grinding pans, also known as pan mills, dry pans or by the proprietary name of Incla pans.

Here heavy rotating wheels pulverize and reduce the material to a diameter of about 2 mm. Each grinding circuit yields around 200 tonnes, resulting in a total of 800 tonnes from the plant's four grinding pans.

After leaving the grinding pans, yet another conveyor carries the material to a BIVITEC sieving screen which eliminates any remaining larger-sized particles.

13

14

15 The 4 mm metal BIVITEC screen mesh can process 108 tonnes per hour.

16 Sieved clay emerging from the BIVITEC

17 Fine material created during the sieving process is then taken up by conveyor to large metal clay storage bins (eight bins in total), each capable of holding up to 150 tonnes.

18 Outfeed from the 150-tonne capacity clay storage bins, feeding material to the pre-mixer.

19 These units are known as loss in weight (LIW) feeders. They work automatically to dose the exact percentage of manganese onto the ground clay before it enters the pre-mixer. Percentages vary by product and each product has a unique recipe selected by the operators on their control panels. External silos are used for bulk storage of manganese oxide.

20 The pre-mixer is fitted with a rotating shaft with blade extensions that act like an auger, propelling the clay (which now has a moisture content of about 8% by volume) on to a conveyor.

Pre-mixing A controlled flow of fine clay particles exits the storage bins and travels to the first pugging mill which 'pre-mixes' the clay. The material is gravity-fed into the chamber of the mill where various substances are introduced by means of a spray system.

The substances include a small quantity of water, liquid sugar, and manganese oxide, all of which help to achieve the desired plasticity for successful extrusion of the so-called 'wet' brick (also known as green bricks, i.e. unfired). The choice of additive depends on the specific type of bricks being produced. Manganese oxide, with its distinct blue-black color, is a commonly used additive.

19

18

20

Secondary pugging, de-airing The journey continues to the second pugging mill located inside the main plant. As the clay travels along the conveyor, moisture continues to evaporate.

The dirt is gravity fed into the mill and worked for one final time to achieve a blended and uniform clay mass with a pliable consistency.

21 External water storage tanks.

22 Powerful belt-driven machinery relays power to the pug mill.

23 Rotating blades in the pug force the clay into a vacuum chamber where any trapped air bubbles are removed through a de-airing process. This improves the workability and plasticity of the clay and the strength of the fired brick.

24 Spare parts for the auger and assembly casing.

25 The clay emerges in one continuous column, often referred to as a 'slug,' and characterized by identifiable perforations. The holes created as the compressed clay is pushed through the die reduce the brick's weight and contribute to its structural integrity and strength.

25

A stream of clay is forced out of the extruder by an auger connected to a rotating shaft.

24

Stage 5: The extrusion process

The clay's journey from the second pug mill, then through the vacuum chamber and into the extruder is a continuous process.

A stream of clay is forced out of the extruder by an auger connected to a rotating shaft. The auger's winding motion forces the clay through a cone-shaped tube oriented horizontally. This tube tapers down to a die with multiple metal prongs, forming a shape resembling that of a brick.

The JC Steele 120 extruder plays a pivotal role in the story of Plant 2. Capable of processing 120 tonnes of clay per hour, it actually exceeds the plant's current capabilities. This groundbreaking technology, crafted solely for Brickworks, is a global innovation.

In fact, the extruder's output corresponds to crafting 40,000 standard bricks within the same timeframe. In other words, it can supply enough material to build five houses every hour.

26 Waste material drops onto a conveyor which returns it to the pre-mixer and the repeat of the extrusion process.

27 Extrusion waste.

28 The die is made from steel and has a number of prongs and is securely attached to the machine's front. It can be replaced with a different die to suit the specific type of extruded brick being manufactured.

29 Depending on the desired brick shape, the front of the die can be fitted with a distinct 'shaper cap,' suitable for round, square, or rectangular bricks. Although some specialized or architectural bricks are made using extruders, these are usually produced through the dry press method at Bowral Bricks in the Southern Highlands, New South Wales.

30 Slugs of clay.

31 Texturing of clay.

32 Wire cutting.

33 Setting robots.

30

Precision cutting of the wet brick is essential and must account for the natural shrinkage that takes place during the drying and firing stages.

31

Stage 6: Forming the bricks and setting on kiln cars

From extruder to cutters The column of clay travels along metal rollers and is kept in a straight alignment by small rollers on either side of the conveyor. At this point the slug can be textured or surface coatings applied to create the desired effect on the surface of the brick. The clay arrives at a transfer point where it pauses briefly before being cut by a guillotine into smaller lengths ready for the next stage. The columns are pushed into position where an automatic cutter resembling a wire frame slices through the column. This process creates the individual brick shapes. Precision cutting of the wet brick is essential and must account for the natural shrinkage that takes place during the drying and firing stages.

32

33

34

The length of time unfired bricks spend in the dryer depends on the 'push rate' of the firing kiln.

35

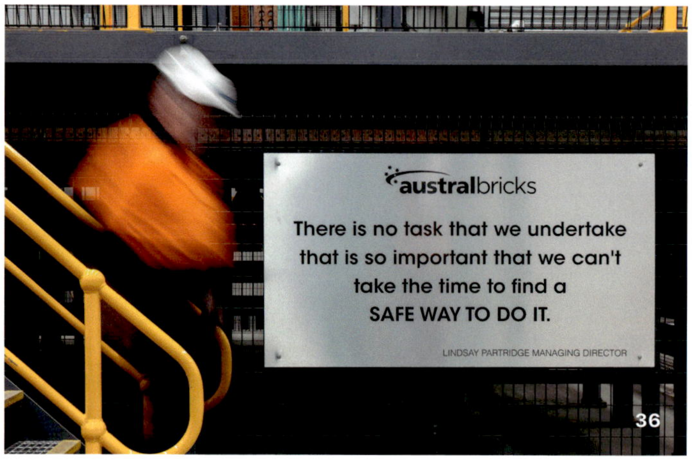

36

37

34 Kiln cars before being set with wet bricks.

35 Pugging and setting are monitored from a control desk at every station.

36 Safety is of paramount concern.

37 Wet or unfired brick

38 Heat extracted from the firing kilns during the firing process is recycled to circulate in the dryers. This dual-use of heat optimizes thermal efficiency for both drying and firing operations. At this stage the combination of heat and humidity is carefully regulated to prevent any risk of cracks developing in the bricks.

39 Plant 2 incorporates three high-capacity dryers, a design choice that contributes to the elongated dimensions of both the dryers and the kilns. Bricks exiting one of the three drying kilns.

40 The Portique kiln is a steel frame lined with fire bricks. It measures 200 meters in length and 6.5 meters in width and is fitted with automatic gas burners and a convective heat exchange system.

41 Firing is monitored from a central command room.

38

39

Setting on kiln cars At the setting station, a robotic arm gently transfers the bricks onto a kiln car prior to the drying phase. The cars move slowly along rails to the drying kilns. Kiln cars are equipped with heat-resistant silica bricks and move along rail tracks. The quantity of bricks on each kiln car is contingent upon the kiln's size. Specifically, at Plant 2, a single kiln car accommodates 7776 bricks, equivalent to the amount needed to construct a whole house.

Stage 7: Drying the unfired bricks

After being set on the kiln cars, the unfired bricks move along the rail track to enter a drying kiln where excess moisture is removed. The majority of the moisture is evaporated within the dryer, where temperature range from approximately 38 °C to 204 °C.

The length of time unfired bricks spend in the dryer depends on the 'push rate' of the firing kiln. A swifter movement of kiln cars laden with fired bricks translates to a shorter period that green bricks need to remain within the dryer.

40

41

42

43

44

Stage 8: Firing

Firing process After exiting the dryer, the loaded kiln cars progress into the tunnel kiln where they will move through a series of 19 zones involving heating and cooling. The continuous burn process gradually moves forty-five kiln cars, each laden with unfired bricks.

Once inside the kiln, the loaded cars roll past stationary fires powered by gas and injected from above and alongside. The heat increases in intensity, all the while being continuously monitored and adjusted by operators using touch screens and computer technology.

The firing rate is carefully managed by a computer supervision system which can be accessed directly from any location around the world.

Cooling phases The cooling phases begin as soon as the cars transition into the cooling zone within the kiln, and lasts for less than 10 hours. This stage significantly influences the final color of the bricks.

Kiln care Ensuring the kiln's upkeep is of utmost importance, however, once it's ignited, no access can be granted. It's worth noting that a tunnel kiln remains continuously operational, sometimes running for a span of 10 years or longer. Regular maintenance of the kiln cars is also carried out to ensure smooth operations.

Stage 9: Dehacking and packing

After exiting the kiln, the rail cars advance to a designated area where the bricks are allowed to cool down completely. Bricks are then placed onto the so-called dehacking system. Using robotic arms, the benches or rows of bricks are gently lifted off the kiln cars and they are systematically stacked into packs. Each pack typically contains 400 bricks, although there's an option for 520 bricks per pack, particularly for interstate transport. In times past dehacking was performed by manual labor.

As successive layers of bricks are placed one on top of the other for a pack, plastic 'time boards' are strategically placed at intervals by the robotic arm to reduce potential damage during transit. The brick pack is ultimately secured using plastic ties or straps, effectively holding it together. The configuration of these packs is designed to facilitate handling by forklifts outside in the yard. Brick packs inside the factory continue by conveyor to emerge outside the factory.

46

42 Firing requires a deliberate and slow progression to avoid inconsistencies in brick firing. Within the firing zone, temperatures are elevated to a range between 1,000 and 1,200 degrees Celsius. To illustrate, it is rather like baking a cake, except that the kiln is designed to withstand extremely high temperatures. Unlike metal, clay gradually softens and vitrifies as it is exposed to increasing temperatures.

43 View down the side of the kiln.

44 Once the desired temperature is reached, the kiln cars carrying the fired bricks exit the tunnel. At Plant 2, more than 40 metric tonnes of bricks exit the kiln every hour, an output sufficient for constructing two suburban houses.

45 Exhaust stack and scrubbers which removes toxic fluoride gases released from the clay during the firing process. Clean air is vented through the stack into the atmosphere.

46 Cooling fired bricks.

47 Dehacking and setting on pallets.

45

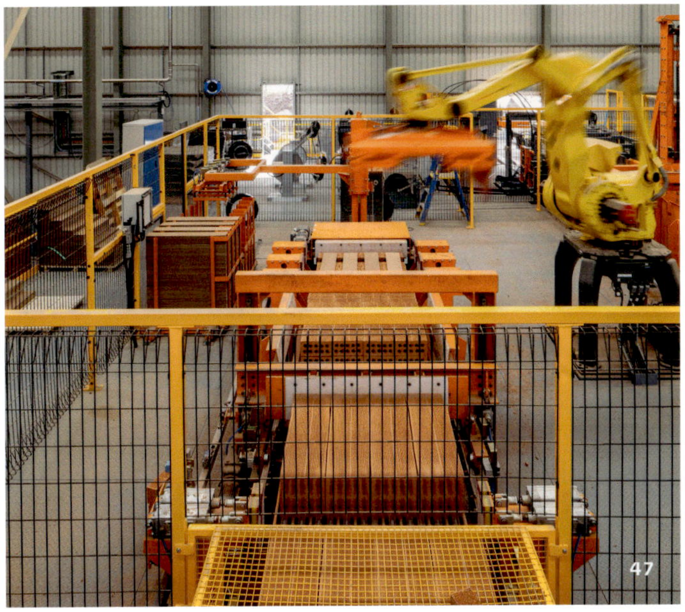
47

48 Brick packs exit factory.
49 Truck collecting brick packs.
50 Bricks in yard.

49

48

Stage 10: Storage yard, ready for delivery

The securely strapped brick packs emerge into the open air on rollers. These carry the bricks to where they can be collected by forklifts. Outdoors in the storage yard, forklifts methodically stack packs, one on top of the other. To accommodate the substantial volume of bricks managed by the robotic dehackers, Brickworks procured three 16-tonne forklifts, each capable of lifting three packs simultaneously. Brick packs are transported to the construction site via trucks equipped with cranes or specialized forklift vehicles capable of navigating challenging terrains.

50

What happens during the firing process?

Stages	Temperature range °C	Description of reactions during firing
Evaporation	20-150	Water-smoking is a process where clay is slowly heated before shaping it. This helps to remove any water mixed in the clay. Gradually increasing the temperature in this stage is important to avoid cracking of the bricks. This is because the outer surface of the bricks contracts faster than the inside when heated quickly. As the water evaporates, it causes a cooling effect, resulting in a drop in temperature in the surrounding area. This whole process is called an endothermic reaction.
Dehydration	149-650	During the firing process, carbonaceous matter, carbonates, and the water contained within the clay are burned off and broken down. The specific temperature at which these substances are fully combusted depends on how fast the clay is heated. It's important to avoid rapid heating, as it can lead to the presence of steam around and inside the bricks. This quick heating can cause a lack of oxygen within the bricks, resulting in undesirable effects like discoloration, dark coloring, and the formation of bloated or cored bricks. As the firing continues, an endothermic reaction occurs because more water and carbonaceous matter are released from the bricks.
Oxidation	300-982	To make high-quality bricks, it's important to completely burn off any leftover carbonaceous materials and convert all iron residues into their oxide forms during this part of the process. The oxidation process within the kiln starts at a certain temperature, which is influenced by factors like the heating speed, the level of carbon in the clay, the quantity of extra air in the combustion area, and the density and shape of the clay bricks.
Vitrification	900-1316	Vitrification is the process that transforms clay into a sturdy, solid structure with relatively low absorption of moisture. This transformation usually begins around 900 °C, once all the carbon-containing materials have fully oxidized, and continues until the highest temperature the bricks can endure without harm is reached. The strength of the fired bricks develops during vitrification, as the clay particles fuse together through sintering and melting. As the clay mass melts, it coats the solid particles and upon cooling, solidifies into a glass-like substance, binding the particles together. The strength of the fired bricks is influenced by factors like the peak temperature achieved, the duration of the vitrification phase or the time at the maximum temperature, as well as the presence of certain fluxes in the clay, like potash, soda, magnesia, lime, and ferrous oxide. During this stage, a series of heat-releasing reactions occur due to the gradual oxidation of sulfur compounds and any remaining organic materials. Additionally, new crystalline structures may form as part of these changes.
Flashing or reduction firing	1150-1316	The highest temperature, also known as the finishing temperature, is maintained for a certain time to achieve the desired color of the bricks. This is done by introducing 'un-combusted fuel' into the kiln.
Cooling	1316-20	After reaching its highest point, the kiln's temperature is gradually lowered back to the normal room temperature over the course of several days.

Source: Akinshipe O, Kornelius G (2017) Chemical and Thermodynamic Processes in *Clay Brick Firing Technologies and Associated Atmospheric Emissions Metrics-A Review*. J Pollut Eff Cont 5: 190. doi: 10.4176/2375-4397.1000190

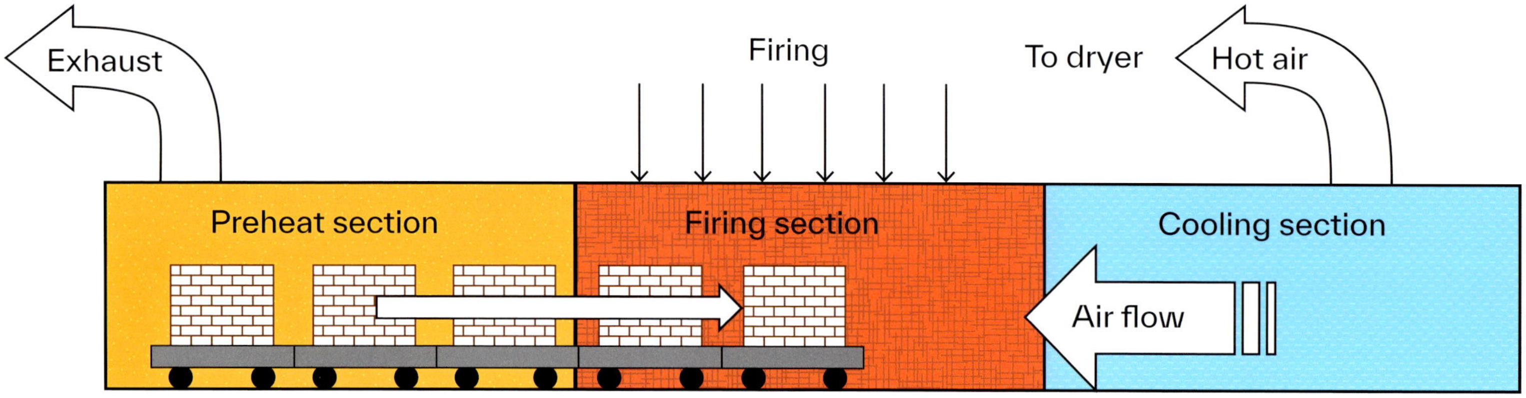

Schematic diagram of tunnel kiln

Quality control Conducting compression tests on every single unit in a batch is not feasible, given the large quantity of units typically involved. Instead, only a small subset of samples is tested relative to the total number of units. To address this, a standard has been established using the 95% characteristic value at a 75% confidence level. This implies that there's a 75% level of confidence that the strength of 95% of the units in the batch is greater than the characteristic strength derived from testing the samples.

Batch color consistency is a notable feature. Thanks to modern kiln technology, everything is made uniformly. This technology's advantage lies in our ability to ensure consistent colors and have control over the composition of each batch.

Australia Standards—*An Introduction to the Australian Standard AS 3700* which outlines the specifications for designing and constructing masonry structures.

Chemical composition of clay Silica and alumina, essential components in brickmaking, are found in a variety of clay materials. The primary percentages are approximately:

- Silica (sand), making up about 50% to 60% of the weight.
- Alumina (clay), constituting around 20% to 30% of the weight.

The durability of bricks significantly relies on the correct balance between silica and alumina. A higher proportion of silica results in more well-shaped and uniformly textured bricks. Silica plays a role in preventing raw bricks from cracking, shrinking, and warping. However, excessive silica disrupts cohesion among the clay particles, yielding brittle and weak bricks.

During kiln firing, silica (sand) and alumina (clay) combine to create a significant mineral known as mullite, which is durable and resilient. Chemically, this compound is referred to as $3AlO3\text{-}2SiO2$. Under a microscope, it appears as elongated, slender crystals that function similar to the binding straw in adobe, interlocking to hold the mixture together.

Iron is a minor ingredient that oxidizes to form hematite when subjected to high temperatures. This iron oxide is responsible for the characteristic red color of many bricks.

Additional elements such as sodium, manganese, calcium, potassium, and lime (comprising 2-5% of the weight) aid in facilitating the melting of silica by acting as a flux. These elements are naturally present in various clay deposits. Manufacturers maintain consistency in chemical composition and physical properties by blending clays from different sources and locations within the pit.

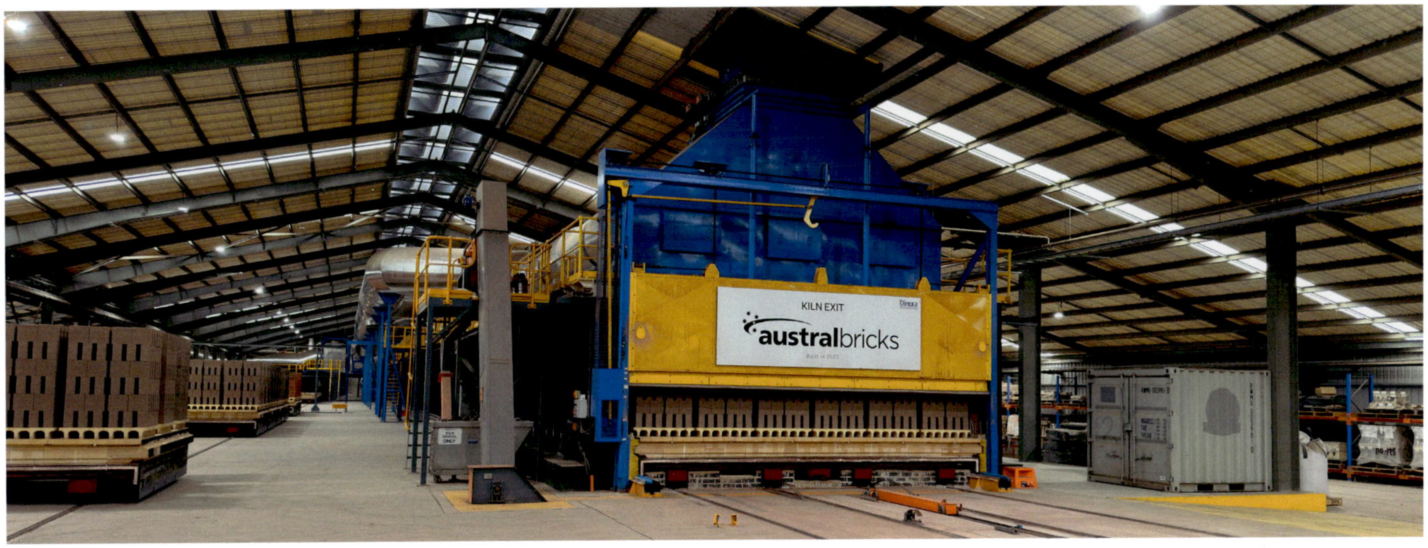

Sustainability at Plant 2

Plant 2 has been meticulously designed and constructed with a strong emphasis on sustainability. The sustainability measures in place for Plant 2 encompass the following features:

- Five rainwater tanks, capable of holding a total of one million liters, are employed for the extrusion process. The plant has the capability to utilize water from the basin dam when tank water levels are low. Alongside basin and tank resources, the plant can also access mains water as needed. Five hydraulic tanks are integrated into the kiln for car propulsion. These tanks operate on-demand, resulting in energy savings.
- The plant operates entirely on electricity, utilizing Variable Speed Drives (VSD), soft starters, and robotics as part of its automation. Soft starting equipment significantly reduces the initial power surge required to activate machinery. LED lights illuminate the entire facility, and motion sensors ensure lights turn off automatically when no movement is detected.
- All clay is recycled throughout the entire manufacturing process.
- The kiln is designed for high gas efficiency, with controlled burners that are utilized only when necessary during the firing process.

Why brick? Brick has many advantages over alternative construction materials that serve the same purpose.

- A durable material, needs little maintenance, and can last hundreds or thousands of years.
- Fireproof and can withstand exposure to high temperatures.
- Offers good noise reduction and sound isolation capabilities.
- Does not need to be painted or coated to protect it from the environment.
- A modular building component so that problems with individual bricks can be addressed without the need to tear down and rebuild the entire structure.
- Made from clay and shale which is available almost everywhere and can be made locally, eliminating the costs associated with shipment.
- Construction using brick as a material may be less expensive than with using stone, concrete, or steel.
- Simpler to work with because of its uniformity in size.
- Can be reclaimed by recycling, starting at the crushing stage.

All images Alex Mayes.

How to use brick to create unique masonry designs

Tim Leese

While brick masonry is one of the oldest construction techniques still used today, its versatility, beauty, sustainability and durability have ensured its popularity through the years. Over the centuries, the use of brick has evolved to create a wide breadth of unique designs that balance functionality and artistry.

Brick's many distinguishing details and available shapes, textures and finishes give architects, builders and designers nearly endless ways to create impactful patterns on the exterior or interior of a home. Brick lends itself to numerous design profiles because it can be bonded in a variety of texture patterns, it can be molded or finished to fit specific space needs, it's available in a range of shapes, and even the mortar itself can be incorporated into the palette.

From historical structures to modern homes, brick is a versatile design element that can act as the foundation or the focal point (or both!), depending on how it is used.

Popular texture patterns with brick

Brick can be used to create texture patterns that are both tactile and visual through different bonding methods. Brick bonding, or the pattern in which the bricks are laid, functions to distribute weight throughout the structure but can also exhibit creative design.

To use texture effectively, first, define the largest source of texture. Too much texture can create a busy, crowded look. For example, a brick accent wall, fireplace, exterior column or other point of convergence could serve as your largest source of texture. Depending on the size of the project, one-to-two distinct texture patterns should suffice.

The following bonding methods each offer a balance of functionality and form and can give a project a distinct look.

Running bond: the most popular bonding pattern, used for basic wall construction. It consists of stretchers offset by ½ brick per course. (Think of your standard brick wall.)

Flemish bond: does not alternate layers but instead shifts them. Each row is completed with alternating full-sized and header bricks. This slight deviation from the running bond adds more visual interest.

English bond: layered with running bond, the contrast occurs with alternating courses of headers and stretchers.

Common bond: similar to the English or running bond, but incorporates headers inserted every five to six layers.

Stacked bond: places emphasis on the vertical aspect of masonry, this bond can be created with either stretcher or header bricks. Stacked bonds would typically be used in a nonstructural capacity.

Herringbone bond: resembles a zigzag pattern. This decorative type of raking bond conceals layers of diagonally laid brick to bond exposed bricks to the wall.

How to use brick detailing

Brick detailing refers to styles of patterns used to arrange bricks around other design elements. Architects, designers and builders may incorporate brick details around windows, doorways or as an archway. Complementary brick detailing adds visual interest while maintaining a cohesive look.

The following brick detailing styles are commonly used to accent homes.

Soldier course and rowlock: 'soldier course' refers to bricks laid standing up with the narrow edge exposed. Imagine soldiers standing at attention. A rowlock is a course of brick laid on its long side with the short end of the brick exposed. You may find soldier course and rowlock detailing along a window or door sill.

Quoin: a decorative style used to accentuate the corner of a building with short side header bricks and long side stretcher bricks. This design can also be achieved with stone blocks to add more texture and contrast. Expect to see quoins protruding around the corners of a home or on decorative columns.

Corbel: this complex style adds dimension by projecting out from a wall. Corbels are also built to bear weight and may be used with features like balconies. Corbel detailing often appears in corbelled arches, decorative curved elements in the corners of an opening.

Types of brick shapes

Brick is available in an extensive selection of shapes like extruded, molded and handmade brick to compose unique architectural elements. The diversity of brick shapes allows architects, designers and builders to use brick in both functional and aesthetic roles.

Here are some of the common shapes of bricks and how these shapes are typically used.

Flat arches: use bricks of the same size and shape with parallel sides. The voussoirs are designed with a 70° skew-back angle and ⅜" nominal mortar joint.

Circular arches: arches manufactured to fit the type of arch, rise, dimension of the masonry opening, thickness of the mortar joint, and radii of the arch.

Water tables: designed to reduce water penetration but also lend themselves to decorative design details. Water tables can be used to add dimension to flat walls, rotate to form door and window jambs, or can be inverted to create cornice brickwork.

Sills, copings and treads: also designed to reduce water penetration. Sill bricks feature one rounded side, tread bricks are slimmer with one rounded side and coping bricks feature an angular side typically used for the top of a structure. Sills, copings and treads can be used on wall caps, windows and door openings.

Mortar as a design element

Mortar is needed for the practical purpose of binding bricks together, but it can also become an important design element by adding a subtle or definitive accent.

Use mortar as a subtle addition or definitive accent in any of the following ways.

Mortar color: standard mortar is usually a shade of gray, but coloring can be added.

Match or contrast: just like with different shades of brick, different shades of mortar can be used to either match or contrast. For example, black mortar can be used to create a dramatic outline with red or gray brick.

Play up single colors: when working with a multicolored pattern of bricks, mortar can be used to underline the primary color. Use mortar to make a wall look lighter or darker, warmer or cooler.

Accentuate patterns: mortar can also mirror greater architectural patterns. For example, with quoins, bands or other ornamental elements, matching mortar can create an even greater visual impact.

Mortar can also become an important design element by adding a subtle or definitive accent.

Mortar joints: defined as the filled space between bricks, can be applied in a variety of ways to complement or contrast other patterns.

Raked joints, flat mortar joints, slightly recessed from the front edge of the brick, and flush joints, mortar that is directly level with the brick, are great options for showcasing the mortar traditionally.

For a more textured look, you may opt for concave or groove joints, vee or v joints, or grapevine joints. These indented techniques draw more attention to the joints.

To maximize your mortar visibility, try extruded or squeezed joints, or overmortar or 'German Smear'. These styles place emphasis on the mortar as a design element and make it a focal point of the project.

Approximately 20 percent of the surface of any brick wall is mortar, do not overlook this potential canvas!

Approximately 20 percent of the surface of any brick wall is mortar, do not overlook this potential canvas.

Conclusion

Brick can be used to create many unique and stunning designs both on the interior or the exterior of a building serving as a canvas for both the architect and the mason's creativity, the balance of form and function, science and art. From the brick itself to the way it's bonded, mortared and stacked, architects, designers and builders have numerous ways to showcase new patterns while paying homage to one of the most enduring construction materials of all time.

Different brick bonding methods can produce different texture profiles, even within the same project. Brick detailing can be used to accentuate focal points of the project. Varying brick shapes within a project can create subtle texture at points of convergence throughout a project. The mortar color and application technique itself offer another palette to complement and contrast the brick itself.

Article by Tim Leese, Director of Marketing, Glen-Gery, first appeared in MASONRY Magazine, *published by the Mason Contractors Association of America (MCAA), the national trade association representing mason contractors. It has been reproduced in full, courtesy of MCAA and Glen-Gery.*

Varying brick shapes within a project can create subtle texture at points of convergence throughout a project.

Brick masonry bonds

Ron Ringer

The rules governing brick bonding have remained unaltered for nearly two centuries.

Compressive and flexural strength in walls

The significance of compressive and flexural strength in walls is underscored by the bond type employed in brick masonry. The bond utilized in brick masonry serves a crucial purpose in achieving the following objectives:

- disrupting the continuity of vertical joints across successive brick courses
- enabling the structure to function as a unified mass
- providing an aesthetically pleasing appearance through the arrangement of bricks in a specific pattern.

Numerous factors contribute to fortifying the strength of a brick wall, including:

- material type and its inherent properties
- workmanship during the construction process
- environmental conditions post-construction
- bond strength
- moisture content percentage in the brick during installation
- wall age.

Note: In this feature article, the 'walls' referred to are mostly solid type walls with essentially a double row of bricks rather than a single row of bricks, as used typically to provide a facing material to a framed structure, i.e. brick veneer.

Walls constructed using bonded brickwork and other masonry types adhere to predefined patterns due to the standard size of bricks. Crafted to conform to a standardized system, bonded brickwork not only presents a tidy look but also minimizes the necessity for brick cutting, while also ensuring an even distribution of forces exerted on the brickwork by floors, beams, roofs, and other structural elements.

English and Flemish bonds stand out as the most widely employed forms of bonding for achieving structural integrity. Bonds in brick masonry are created through the mortar filling placed between layers of bricks and within the grooves formed when bricks are laid side by side and in tiers within walls. While cement mortar primarily finds usage in brick masonry, lime mortar and mud mortar can also be employed, contingent on the construction type and environmental conditions, i.e. mud mortar is not suitable for use in wet climates.

Bonding is instantly recognizable. Vertical joints amidst the bricks on the wall's surface are 'staggered', meaning they do not align across adjacent courses. This staggered effect is accomplished by arranging headers (the shorter face of the brick) and stretchers (the longer face of the brick) in a pattern, although numerous combinations can be utilized. Vertical joints in alternating courses should be aligned directly above one another, whether on the wall's exterior or interior; in technical terms, these should be maintained as perpends, i.e. a vertical layer of mortar between two bricks.

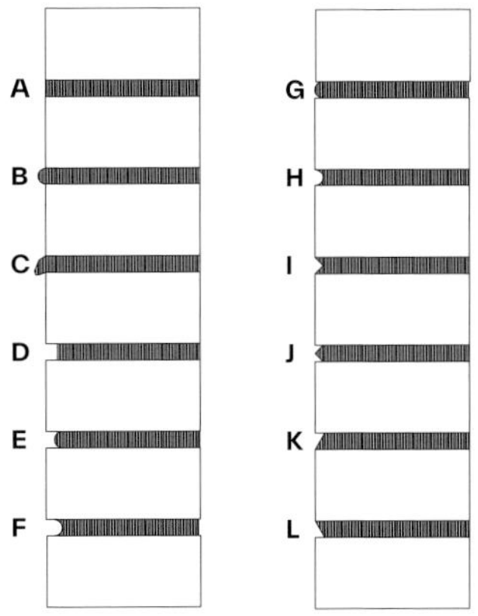

Brick mortar joints

A Flush or flat joint
B Extruded or convex joint
C Squeezed or unfinished joint
D Raked or recessed joint
E Recessed and beaded joint
F Recessed and rodded joint
G Beaded joint
H Concave or rodded joint
I Scored joint or V-joint
J Convex V-joint
K Weathered joint
L Struck joint

Achieving a robust bond necessitates uniform brick size and consistent arrangement throughout the wall. Interior wall bricks should be laid as headers (across the wall) unless specific cases arise, such as footings, corbels, and cornices. Nonetheless, the aforementioned brickwork styles are predominantly employed in the restoration of historic

structures rather than in residential constructions. The rules governing brick bonding have remained unaltered for nearly two centuries, especially since the advent of uniformly shaped dry-pressed bricks in the late nineteenth century.

Structural vs. pattern brick bonds: making walls strong and beautiful

Within the realm of brick masonry, an array of bonding patterns exists, all serving the fundamental purpose of establishing structural integrity. These patterns adapt depending on the type of wall under construction. The term 'bond' in brick masonry encompasses two meanings, which collaborate rather than oppose each other in practice:

- Structural bond: This form of bond is utilized to erect walls that carry substantial loads, providing them with the necessary load-bearing capacity.
- Pattern bond: Employed in walls that serve both load-bearing and non-load-bearing functions, pattern bonds contribute to aesthetics while retaining strength.

Numerous factors influence the selection of a brick bonding pattern, including required strength, structural thickness, specific location within the structure (e.g., a course of bricks above a window might adopt a distinct pattern), aesthetic considerations, brick type and size, and local traditions.

It is worth noting that the usage of either structural or pattern bonds is not mutually exclusive.

Load-bearing walls

In the context of structural bonding, individual masonry units interlock or intertwine, transforming the assembly into a singular structural entity. Structural bonds can be established through various means, including:

- overlapping (interlocking) masonry units based on adaptations of traditional bonding methods such as English bond and Flemish bond
- incorporating metal ties within connecting joints
- introducing grout into the cavity or collar joint between layers of masonry.

Examples of structural bonds encompass English bond, Flemish bond, Stretcher bond or Running bond, and Common or American bond, each with potential variations. Even the so-called garden wall bonds, like English and Flemish (also known as Sussex), Boundary Wall, Scotch, Common, and Monk (Yorkshire bond) reflect local distinctions for garden wall bond patterns.

All structural bonds can be applied to craft common patterns in both load-bearing and non-load-bearing walls.

Contemporary building codes necessitate that masonry-bonded brick walls adhere to certain standards: headers should comprise approximately four percent of the wall's surface area, and the distance between adjacent headers should not exceed 60 cm vertically or horizontally. When combined with variations in brick color, texture, and joint types, these bonds allow bricklayers to generate an extensive array of patterns.

These patterns can stem from the structural bond used or purely serve decorative purposes, detached from structural elements. In essence, a pattern pertains to the alteration or diverse arrangement of brick texture or color on the facade. This flexibility enables the utilization of various structural bonds to achieve numerous patterns.

Non-load-bearing walls

Pattern bonds emerge as ornamental solutions for both non-structural and structural walls. In instances of freestanding walls or residential buildings where slight strength reduction is acceptable, basic bond patterns are often diversified to increase the number of stretchers on the wall's face. This adjustment reduces the number of bricks in the visible area, consequently lowering the cost of thick wall construction. These modified bonds are often referred to as 'garden wall' bonds, although their application extends beyond garden walls.

Although the term 'garden wall' might suggest a specific context, this type of bonding can be applied to a spectrum of brick walls. In the United States, the use of brick in garden walls has become uncommon, with many walls now constructed from materials like wooden paling, aluminum panels, or faux ironwork fashioned from extruded aluminum tube.

Left Tashkent, Uzbekistan. Image: AXP Photography on Unsplash.

Previous spread Vanderbilt University, Zeppos College, by David M. Schwarz Architects. Image: Steve Hall, Hall+Merrick+McCaugherty.

Next spread Grundtvig's Church, Copenhagen, Denmark. Image: Daniele Colucci on Unsplash.

Brick orientation: positioning for structural and aesthetic impact

A course of bricks is one of several continuous, horizontal layers (or rows). In brick masonry, a course refers to a continuous, horizontal layer of bonded masonry units. Irrespective of how bricks are placed within a wall, the size dimensions (thickness, height, and length) are consistently measured from the stretcher position.

The proper selection of brick orientation contributes both structurally and aesthetically to construction projects. Whether it's for strength, visual appeal, or a combination of both, understanding these orientation options is essential for achieving the desired outcomes.

Header This arrangement involves laying a brick with its ends facing the wall's exterior.

Stretcher As the name suggests, this orientation positions the long, narrow side of the brick outward. This allows for efficient 'stretching' of masonry work into walls or borders. By stacking bricks atop each other in straightforward courses, masons can quickly construct significant structures.

Raking stretcher A variation of the stretcher position where vertical joints are staggered by a quarter or three-quarters of the brick's length. This creates a descending visual movement in the bond, with the possibility of extending the zigzag effect left and right.

Soldier Bricks are laid standing on end with the narrow edge facing outward. Often used for decorative effects above doors, windows, and fireplace facings, this orientation provides a vertical structural line.

Sailor Similar to the soldier position, the sailor brick stands on its short edge but faces outward with the wide edge. While offering a flat, vertical structural line, it requires additional front-to-back support for larger structural builds.

Rowlock The rowlock brick is a soldier orientation that tilts its short, vertical edge outward. This setup supports rows of various orientations in alternating patterns, contributing to overall structural integrity and appealing finished patterns.

Rowlock stretcher Also known as a 'shiner', this orientation tilts the face of the stretcher upward, presenting the long, wide edge outward. Like the regular stretcher, this promotes horizontal progress and vertical growth in wall construction. However, it necessitates extra front-to-back support due to the narrow side's limited weight-bearing capacity.

Bull header A rowlock masonry unit with its bed placed perpendicular to the wall's face.

Bull stretcher A rowlock masonry unit with its bed aligned parallel to the wall's face.

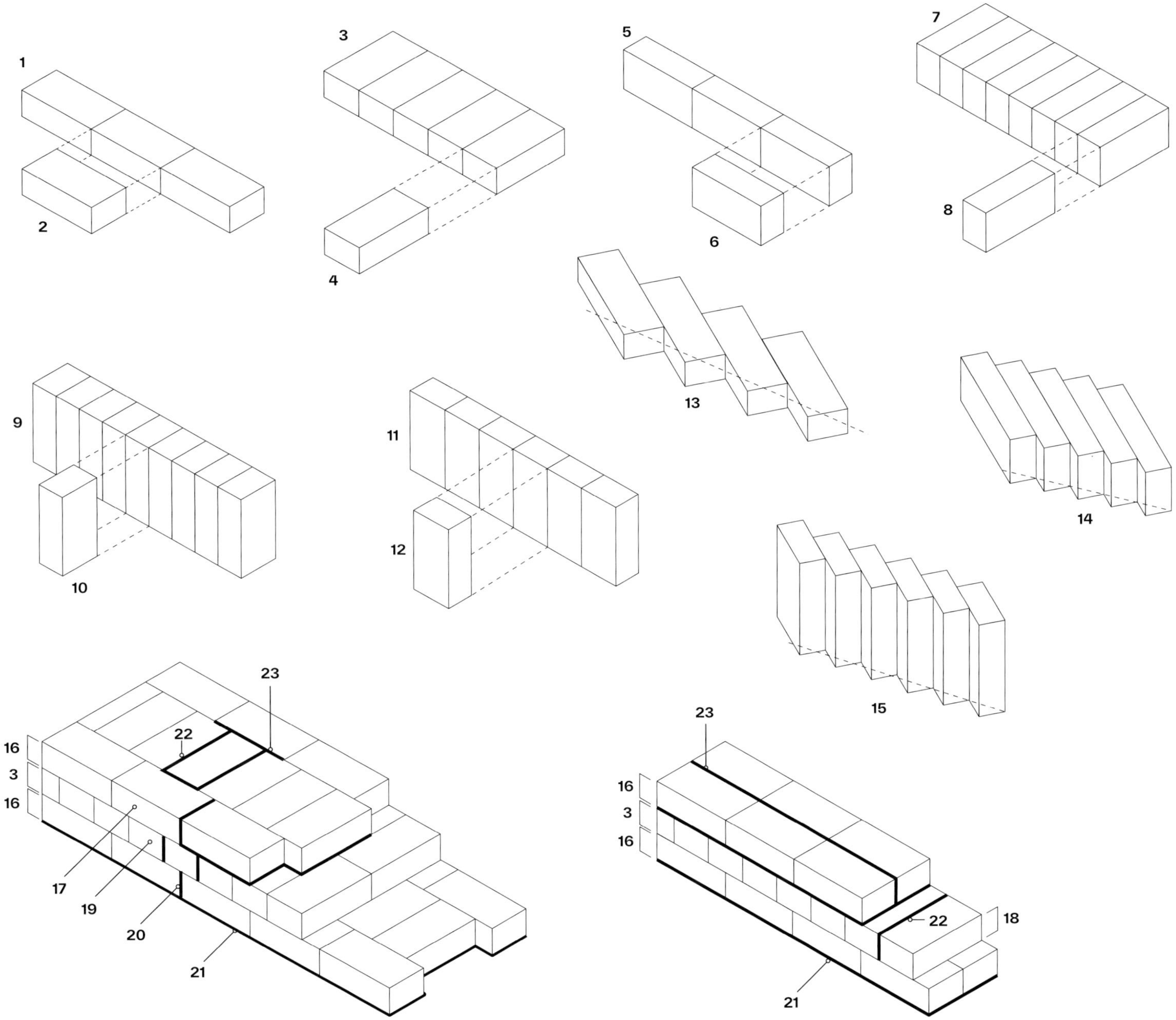

1 Stretcher course, stretching course
2 Stretcher
3 Header course, heading course
4 Header, inbond (scot.)
5 Shiner course, bull stretcher course
6 Shiner, bull stretcher
7 Rowlock course, bull header course
8 Rowlock, bull header
9 Soldier course
10 Soldier
11 Sailor course
12 Sailor
13 Dogtoothed brickwork, bricks on flat
14 Dogtoothed brickwork, bricks on edge
15 Dogtoothed brickwork, bricks on end
16 Stretcher course
17 Stretcher face
18 Binding course, header course
19 Header face
20 Perpend, header joint, perpend joint
21 Bed joint, horizontal joint
22 Cross joint
23 Wall joint

Alterations to Bricks (cuts)

When designing a wall necessitates it, a brick can be divided into smaller segments to suit a specific space. This technique also allows the divided brick to establish an offset or overlap at the commencement of a fresh course. In certain regions such as Australia and the United Kingdom, these divisions were once termed 'bats', though their prevalence was more significant in the twentieth century compared to modern times.

Stretcher bond or running bond

Stretcher bond, commonly referred to as 'running bond', stands out as the most frequently employed bonding pattern. The term 'running' originates from the application of the bonding agent, resulting in a straight alignment between the bricks and other building materials. This bond represents the simplest among the fundamental pattern bonds. It involves the utilization of stretchers, implying that bricks are laid with their long and narrow sides exposed on each course.

Subsequent courses are laid in a half bond (half-brick) manner in relation to one another, excluding return angles and stopped ends. The overlap, typically equivalent to half a brick's width, is achieved by initiating every alternate course with a half-brick bat. Running bond finds extensive use in cavity wall construction, as well as in veneered walls made of brick. It's also frequently applied in facing tile walls, where enhanced bonding can be achieved through the use of wider stretcher tiles. Notably, due to the absence of headers in this bond arrangement, the incorporation of metal ties is often necessary.

Header bond

In a header bond, bricks are positioned with their narrow ends facing outward on the walls' surfaces. This arrangement creates an overlap, typically half the width of a brick. To achieve this overlap, a three-quarter bat is introduced in every other course at the corners of the wall, known as quoins. This type of bond is particularly advantageous for walls with curved designs on the floor plan. However, it finds its primary application in the construction of footings within foundations, serving to enhance the lateral distribution of the load.

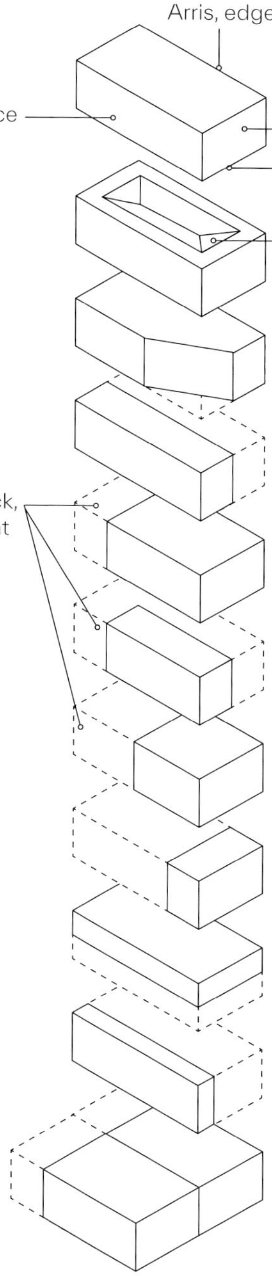

English bond and its variations

English bond, also known as 'ancient bond' in certain European countries, is a brickwork pattern where courses of headers and stretchers alternate. This arrangement represents one of the oldest forms of brick bonding. English bond and its variations were utilized for various brickwork applications until the mid-seventeenth century when Flemish bond was introduced.

As a method of interlocking thicker walls, English bond (along with Flemish bond) provides optimal strength. Nonetheless, it involves significant labor due to its alternating pattern. Several variations exist, involving headers placed every fourth, fifth, or sixth course. Solid brick walls constructed in English bond undoubtedly achieve exceptional strength due to their minimized continuous vertical joints. However, the appearance of English bond tends to be uniform, making it suitable primarily for situations where strength takes precedence over aesthetics.

Rows of headers and stretchers alternating can create a pattern known as 'quarter bond'. In this configuration, headers are centered on the stretchers, and the vertical joints between stretchers align consistently across all courses. Snap headers are integrated into non-structural bonding courses. Achieving and maintaining quarter bond requires positioning a queen-closer adjacent to the corner header. The visual interplay between courses of headers and stretchers accentuates the horizontal aspect. Differing colors or tones between headers and stretchers can result in a striped visual effect.

Variations in layout give rise to subtle cross and diagonal motifs within the brickwork. Introducing extra headers, some of contrasting colors, gives birth to decorative diaper (diamond) patterns that characterize sixteenth-century Tudor brickwork.

English cross bonds As the name suggests, crosses are formed by alternating a course of headers with a course of stretchers, just like in the English bond. There are two variations: English cross bond with king-closer, and English cross bond with queen-closer.

- English cross bond with king-closer: A king-closer is laid at the end of the stretcher course. Every four courses, a header is placed next to the king-closer to create a staggered pattern, shifting by half a brick.
- English cross bond with queen-closer: In this variation, the king-closer is not placed in the stretcher course but in the header course. Moreover, it is positioned not all the way at the end but next to the last header of the course. A header is added every four courses of the stretcher course, resulting in bricks staggered by half a brick.

Koch's bond, Danish bond This specific pattern comprises two whole bricks and a half-brick (known as a half bat). A new feature here is that the end of the half-brick protrudes outward by three centimeters.

Double English cross bond This bond involves placing two courses of stretchers on top of two courses of headers.

Dearne's bond, Hollow Wall bond, Cavity bond Dearne's bond follows a pattern similar to the English bond, but it substitutes stretchers with shiners.

Flemish bond and its variations

Flemish bond stands as one of the most enduring traditional forms of brick bonding, tracing its origins back to the late seventeenth century. The essence of this brick bonding approach involves a rhythmic arrangement of alternating headers and stretchers in each course, where the headers are precisely centered above and below the stretchers. Adjacent to the quoin header, a closer is strategically positioned to establish the accurate quarter bond. The adaptability of this bond becomes evident when adjusting the number of stretchers between headers within each course. For example, a configuration with three stretchers alternating with a header is referred to as the 'garden wall' bond. By combining stretchers of one hue with headers of another, captivating visual effects emerge. The Double Flemish bond presents a replication of the pattern on both the front and rear of the wall, while the Single Flemish bond lacks this rear pattern.

Despite its presence in late Medieval structures dispersed across various regions of northern and central Europe, particularly in Poland, it is noteworthy that this bond doesn't originate from Flanders or its neighboring sections in France and Holland. The term 'Flemish' bond is a product of its connection to contemporaneous edifices in the Low Countries. Unlike utilitarian bonds such as the English bond or garden wall bond, Flemish bond serves as a decorative embellishment, enriching the visual allure of a wall's surface.

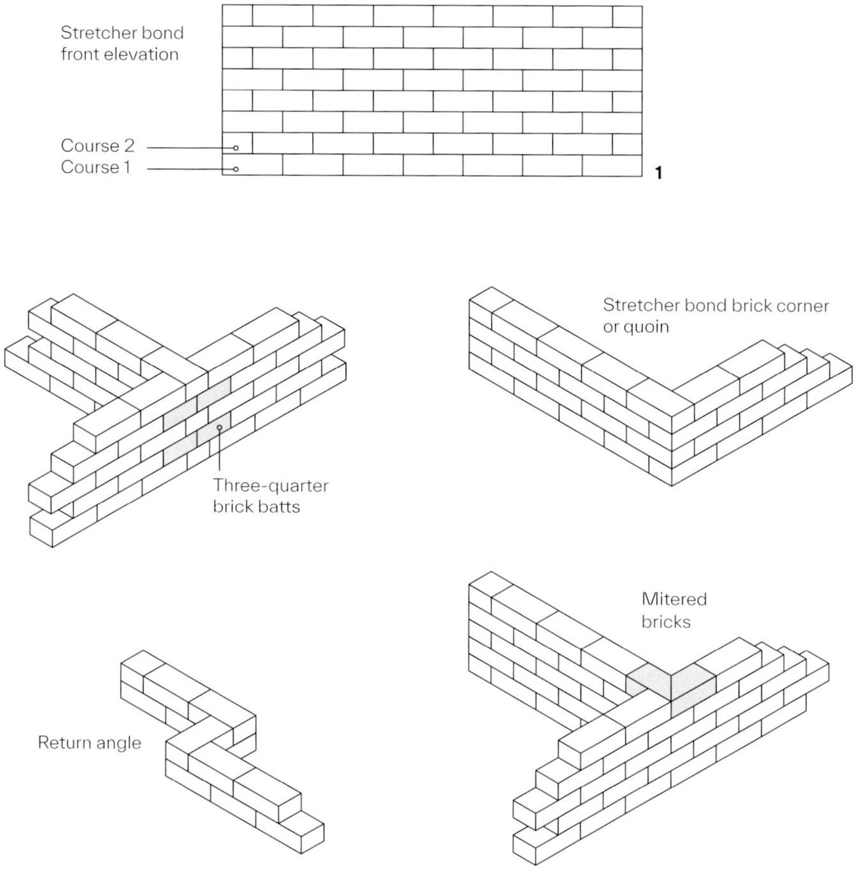

1

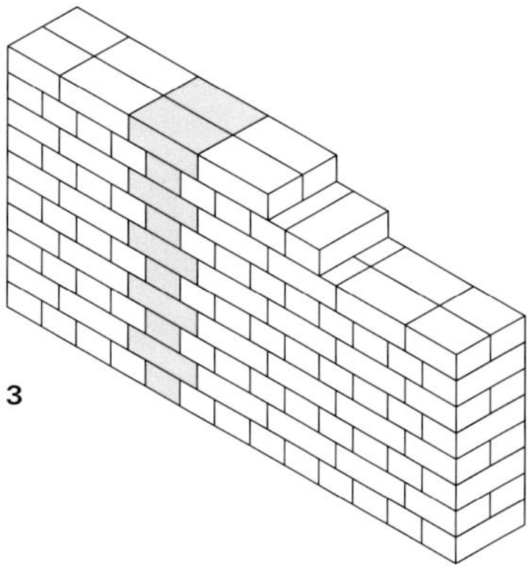

3

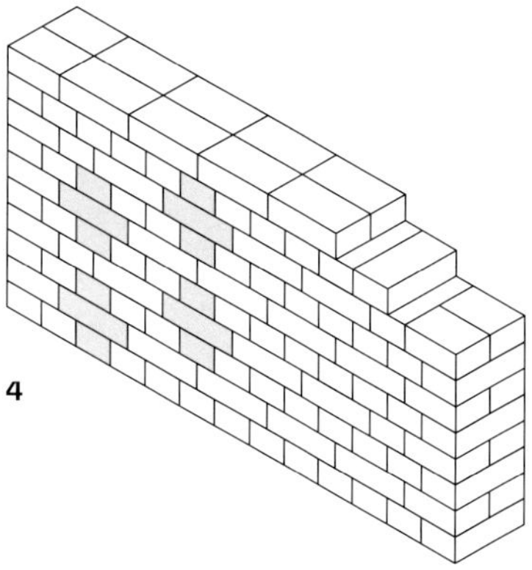

4

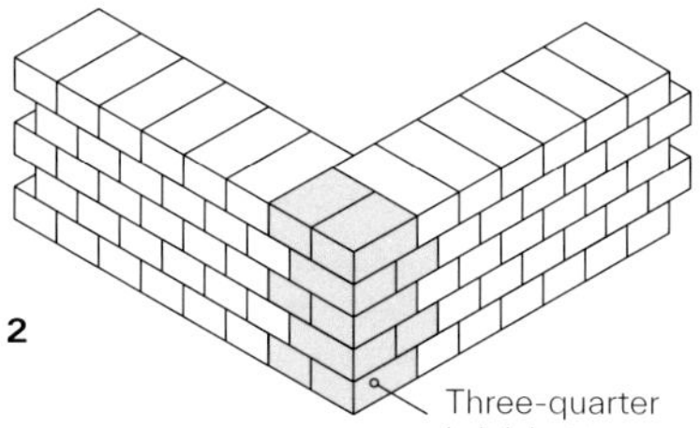

2

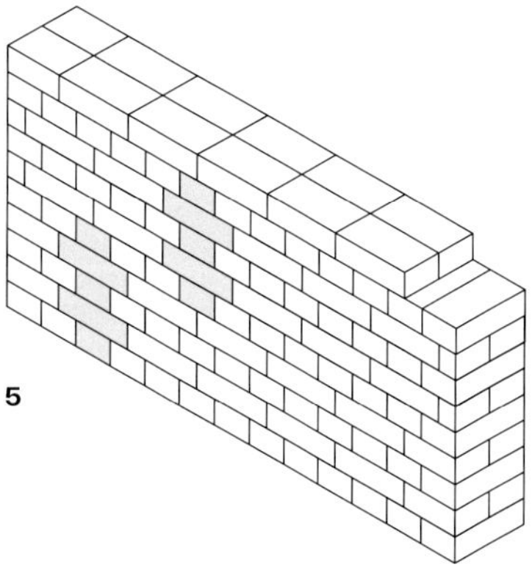

5

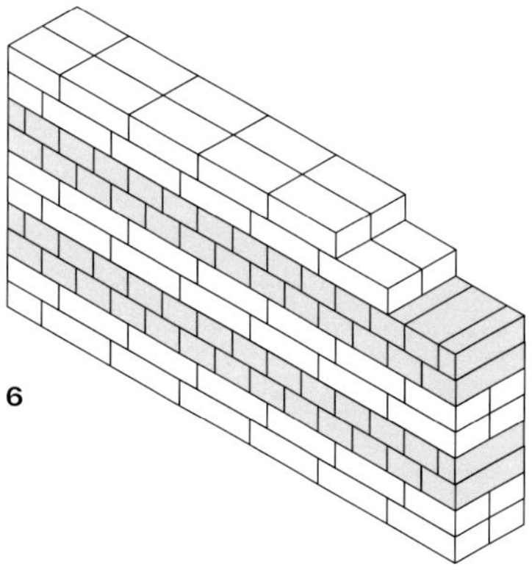

6

1 Stretcher bond or running bond and corner options
2 Header bond
3 English bond with queen closure
4 English cross bond with king closure
5 Koch's bond, Danish bond
6 Double English cross bond

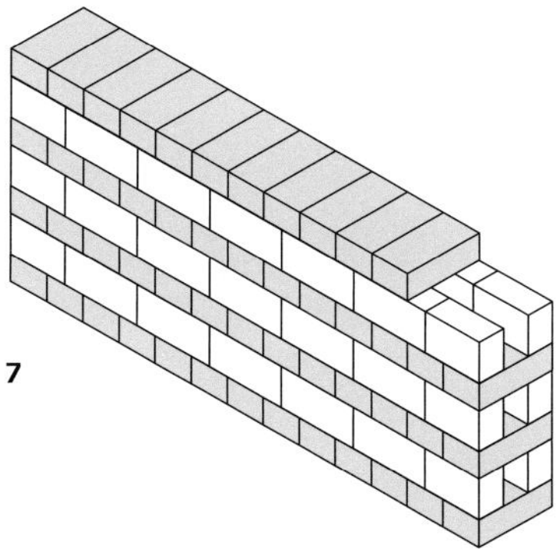

7

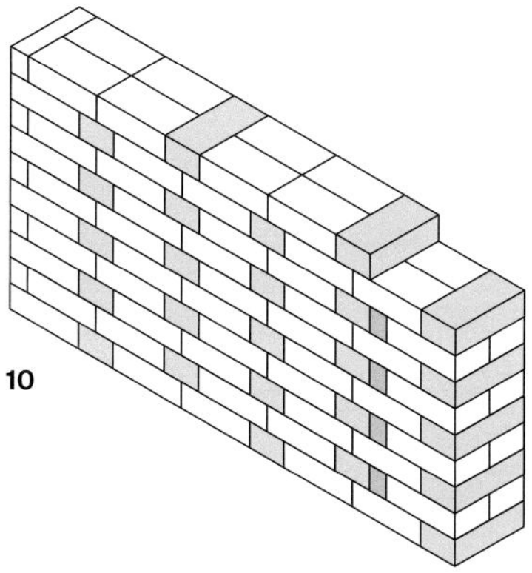

10

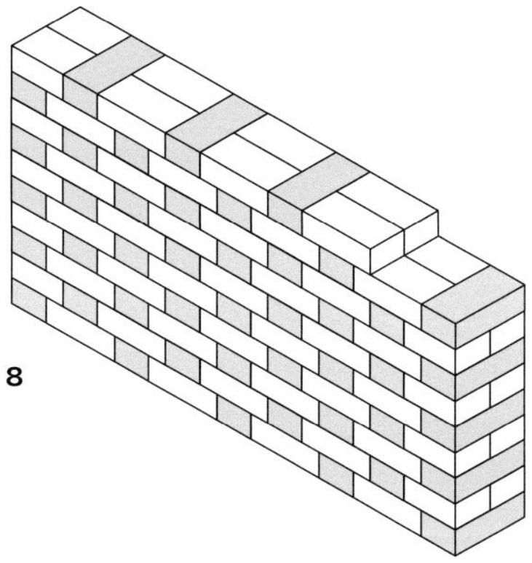

8

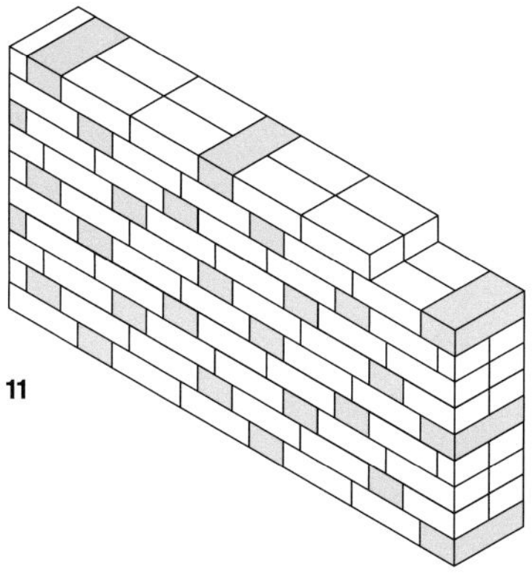

11

9

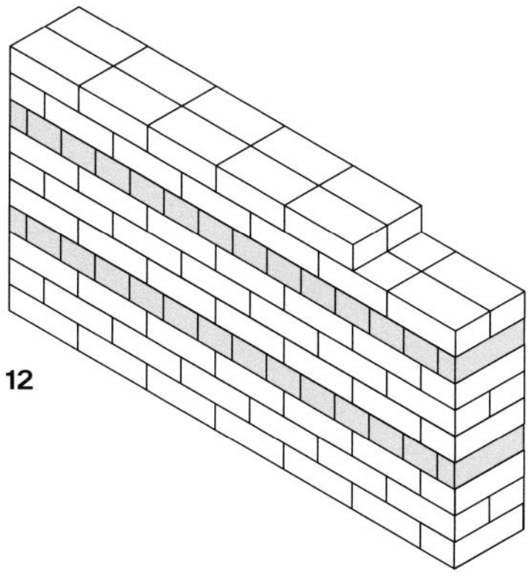

12

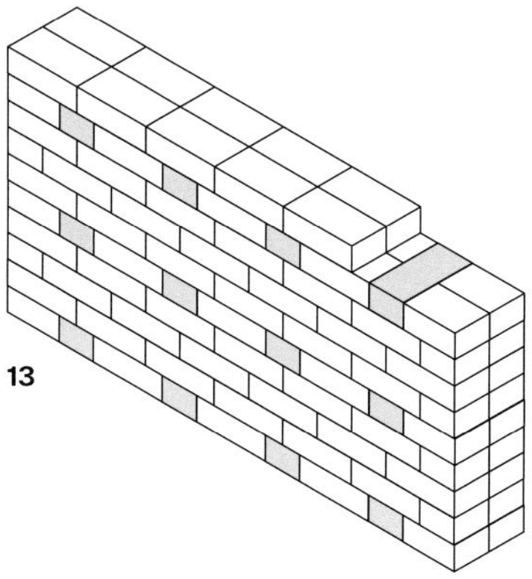

13

7 Dearne's bond, hollow wall bond, cavity bond
8 Flemish bond
9 Double Flemish bond
10 Monk bond, Yorkshire bond, flying bond, Flemish double stretcher bond
11 Flemish double stretcher bond raking variation
12 Common bond, American bond, Liverpool bond
13 Flemish stretcher bond, American with Flemish bond, common bond

The rhythmic interplay of stretchers and headers creates a visually pleasing pattern that necessitates skillful execution. The evolution of Flemish bond in North America occurred during the Federal period, prompted by advancements in manufacturing processes, which consequently led to alterations in the treatment of mortar joints.

In comparison to the English bond, this bonding technique exhibits a slightly weaker structure due to the increased proportion of vertical joints. The prevalence of walls with an odd number of half-bricks in thickness can be attributed to the minimal use of headers within their interiors. Flemish bond proves to be a cost-effective solution for facing bricks, owing to factors such as the reduced requirement for headers per square yard when compared to the English bond, as well as its inherently more pleasing aesthetic appearance.

Double Flemish bond The term double Flemish bond finds its origin in the pattern that graces both the front and rear sides of the wall, creating a mirrored effect. In contrast, when this pattern adorns only the front surface without replication on the rear, the brickwork is dubbed as the Single Flemish bond.

However, it is worth highlighting that the Double Flemish bond, while distinctive, exhibits relatively lower strength when juxtaposed with the English bond. This disparity is primarily due to the higher concentration of vertical joints within its arrangement, and arises from the specific orientation of bricks in the pattern. Furthermore, walls constructed with an odd number of half-bricks in thickness result from the restrained usage of headers within their core structure.

Above Detail of brickwork. Image: Dimitriosfos on iStock by Getty images.

Right M-Museum Leuven, Leuven, Belgium. Image: Bernard Hermant on Unsplash.

In scenarios where there is a preference for premium facing bricks, the use of the Flemish bond emerges as a judicious choice for several compelling reasons:

- Optimal header efficiency: Merely 64 headers per square yard suffice when employing entirely whole bricks within the Double Flemish bond, whereas the English bond necessitates 72 headers per square yard.
- Elevated aesthetics in single-brick walls: When dealing with walls comprising a solitary layer of bricks (termed as a 'skin'), achieving a harmonious and alluring facade on both sides becomes simpler through the application of the Flemish bond, thanks to the inherent nature of its pattern.
- Aesthetic allure: The Flemish bond boasts a general consensus of being more visually captivating than the English bond, thus contributing to its elevated popularity.

Double Flemish bond stems from the symmetrical replication of the pattern on both sides of the wall. While it might not match the strength of the English bond due to an increased prevalence of vertical joints, its cost-effective utilization of high-quality bricks and its aesthetically pleasing demeanor lend it practicality and widespread appeal.

Flare headers and stretchers In brickwork, bricks of a darker color than the main ones are used to create patterns.

Monk bond, Yorkshire bond, Flying bond, Flemish double stretcher bond

A variation of Flemish bond, the Monk bond positions the header directly above the joint between two stretchers. Widely used around the Baltic Sea region until the turn of the thirteenth and fourteenth centuries, it gradually gave way to Flemish bond. The Monk bond offers various arrangements to create staggered courses. For instance, 'raking courses' may be staggered to generate the illusion of diagonal stretcher lines or a subtle appearance of indented pyramid-like diagonals.

The Monk bond displays a diverse array of patterns that allow for flexible layouts at the quoins. Quoin bricks can be stretchers, three-quarter bats, or headers. While the use of queen-closers next to the quoins is optional, these irregular raking patterns pose challenges for bricklayers due to scaffold obstructions, interruptions caused by openings, or other bond-disrupting obstacles. Constructing raking Monk bond masonry can be costly and time-consuming, especially if the bricklayer frequently pauses to ensure correct brick arrangement. Despite these intricacies and associated expenses, Monk bond remains a popular choice for brickwork in northern Europe.

The pattern utilizes intermittent rows of headers, always centered on a stretcher, with the joints of each header course consistently aligned.

Flemish double stretcher bond raking variation

An example of a raking Monk bond.

Common bond, American bond, Liverpool bond

All of these bonds are similar to the Stretcher bond, but distinguished by the insertion of a row of headers every few courses. The pattern utilizes intermittent rows of headers, always centered on a stretcher, with the joints of each header course consistently aligned.

The Common bond provides an efficient means to utilize half-broken bricks, as these half-bricks can be incorporated as header rows without wastage. This bond is particularly suitable for double-thickness walls. Historically, this bond was predominantly found in buildings across the north of England, it is especially prevalent along the east coast. It gained prominence from the late eighteenth century onward and was occasionally employed in garden walls. A 'queen-closer' brick is placed adjacent to the quoin header of the heading course to achieve the necessary overlap.

Flemish stretcher bond, American with Flemish bond, Common bond

Courses of stretcher bond with the insertion of a row of Flemish bond every five to six layers.

Scotch bond

Distinguished by five courses of stretchers between every course of headers, the overlap is achieved by employing headers as quoins for even-numbered stretching courses, counting up from the preceding heading course. Queen-closers serve as the penultimate bricks at both ends of the heading courses.

Garden wall bonds

There are two primary types of garden wall bonds: English garden wall bond and Flemish garden wall bond. In the UK, local names like Boundary wall, Sussex, Scotch, and Common bond refer to the concept of garden wall bonds. Monk bond and its variations can also be classified as types of garden wall bonds.

The brick bond pattern known as the Garden wall bond is essentially a variation of Flemish bond, wherein three stretchers alternate with a header. Alternatively, when there are two stretchers between each header, it is referred to as a 'double stretcher garden wall' bond. Depending on your desired appearance and the wall's purpose, you can also lay four or even five stretchers between the headers.

These variations of the bond are commonly used for constructing one brick thick walls, serving as garden walls or boundaries. The Garden wall bond allows for the creation of uniform wall faces with relative ease and cost-efficiency. However, it's not as structurally robust as English bond and is better suited for constructing dwarf walls or similar structures under minimal stress. Due to its appealing appearance, this bond is occasionally utilized for the outer leaves of cavity walls as well.

English garden wall bond The English garden wall bond involves three rows of stretchers between every row of headers. In the standard application of this bond, headers are used as quoins for the middle stretching course, creating overlap. A queen-closer brick is positioned as the second-to-last brick at the ends of the header courses. For raking wall bonds, achieving overlap requires a more intricate arrangement of quoins and queen-closers.

In some instances, the heading course in this bond features bricks of distinct colors compared to the surrounding stretchers. In chalk-rich regions of England, flint can replace stretchers, and the headers form a lacing course within the bond.

Flemish garden wall bond Flemish garden wall bond is employed for lightly loaded walls, alternating three stretchers with a header. This pattern is staggered in each course. It is also known by other names like Sussex bond and Silesian bond.

In a Flemish garden wall, courses alternate between one header and three (or sometimes even five) stretchers placed in series throughout the course's length. Every alternate course features a three-quarter bat adjacent to the corner header, and a header is placed over the midpoint of each central stretcher.

While headers are infrequently used in this bond, it renders construction in Sussex bond more cost-effective, as bricklayers can progress quickly with sets of three stretchers. Historically, this bond was rarely employed for garden walls and is generally absent in the USA. It's most prevalent in areas like West Sussex and Hampshire in the UK, appearing in around 10% of historic buildings in those regions.

Variations on Flemish garden wall bonds:

- Flemish cross bond
- Raked Flemish bond
- Staggered Flemish bond or Dutch bond
- French bond: every course alternately consists of a stretcher and two headers.

Brick on edge bond

Brick on edge bond, also known as rat-trap, Chinese, rowlock, silverlock, or cavity bond, utilizes stretcher bricks on edge rather than flat. This bond is economically efficient but less strong. i.e. the walls are thinner. Regardless of whether the bricks used are produced by the extruded method or the dry press method, the absence of a 'frog' or indent (where the dry press brick is placed on its wide base, and the indent is filled by an ample amount of mortar) means that these walls are prone to structural weakness and more prone to toppling over. It is commonly used for garden walls and compound walls. Bricks are vertically positioned, forming headers and stretchers to create a continuous cavity, thereby requiring fewer bricks.

Irregular bond, random bond, wild bond

This laying pattern is recommended when mortaring or bonding facing bricks is characterized by low dimensional stability. With reused bricks, there is sometimes no choice.

A random bond uses different lengths of brick, with the exception of dimensions smaller than one quarter of a brick. Each even or uneven row must start with a three-quarter brick. Each other row can start with a header or stretcher. Only headers and stretchers are used in the wall, in any pattern. A maximum of four or five stretchers should be placed next to one another. A maximum of two headers should be placed next to one another. A maximum of five to six steps are possible. Also, this type brickwork is primarily used with joint-free designs, as not all bricks need to be trimmed.

Above Wang Shu's Ningbo Historic Museum, Ningbo, Zheijiang, China. Image: Jadranko Marjanovic Mare.

Following spread Yale University, Pauli Murray College and Benjamin Franklin College, Robert A.M. Stern Architects.

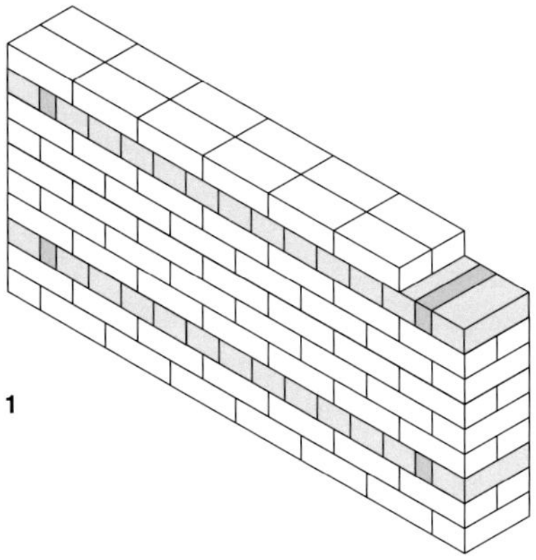

1

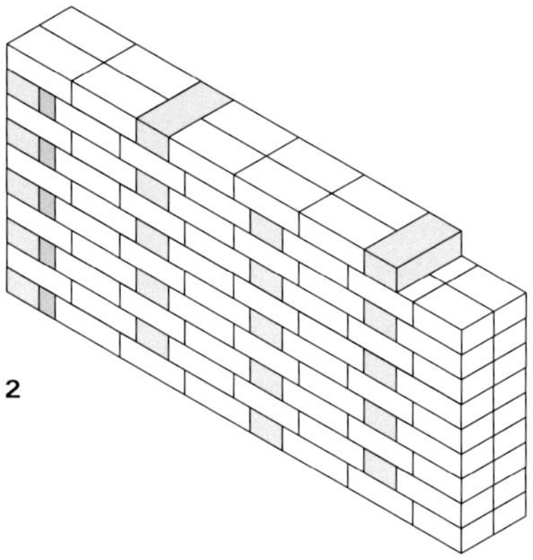

2

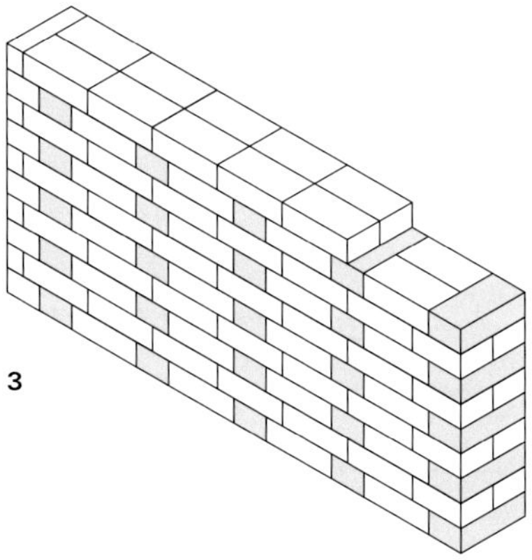

3

1 Scotch bond
2 Flemish garden wall bond
3 Flemish cross bond
4 Raked Flemish bond
5 Staggered Flemish bond or Dutch bond
6 French bond
7 Brick on edge bond
8 Irregular bond, random bond
9 Wild bond
10 Honeycomb-style bonds
11 Stack or block bond

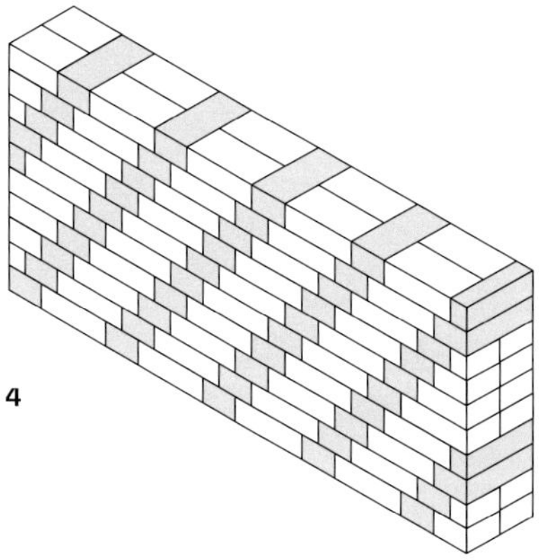

4

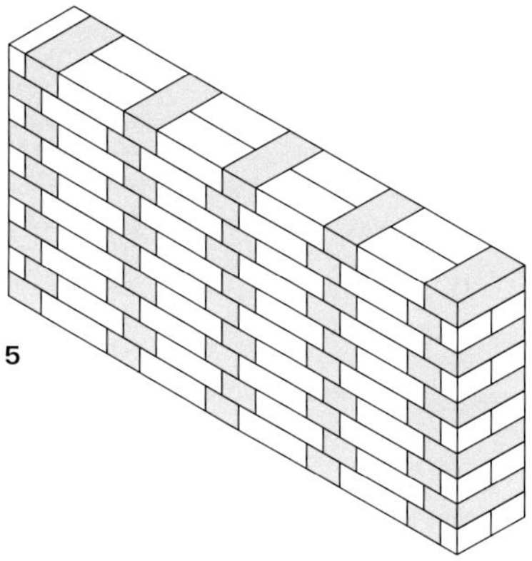

5

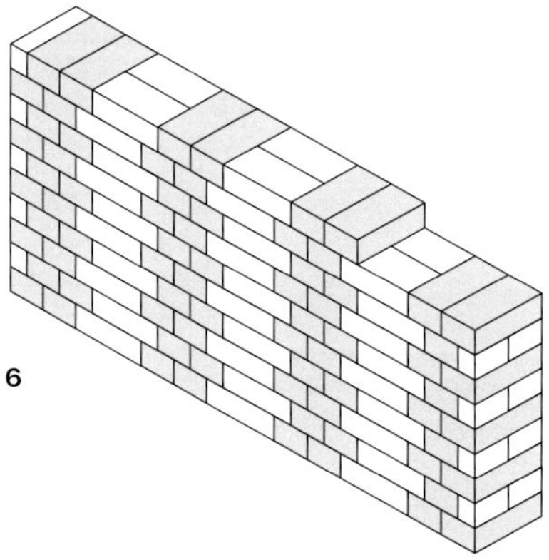

6

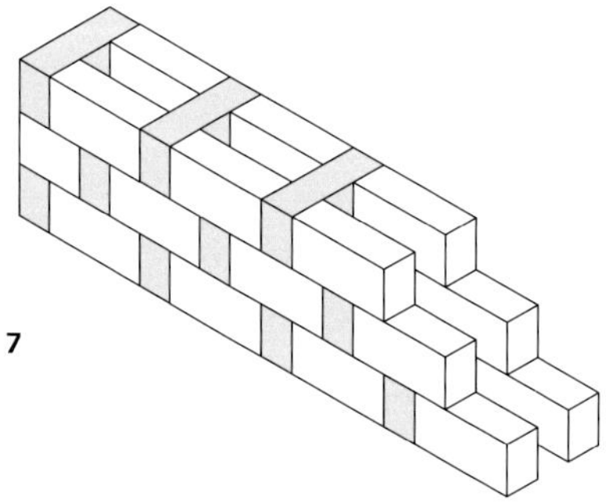

7

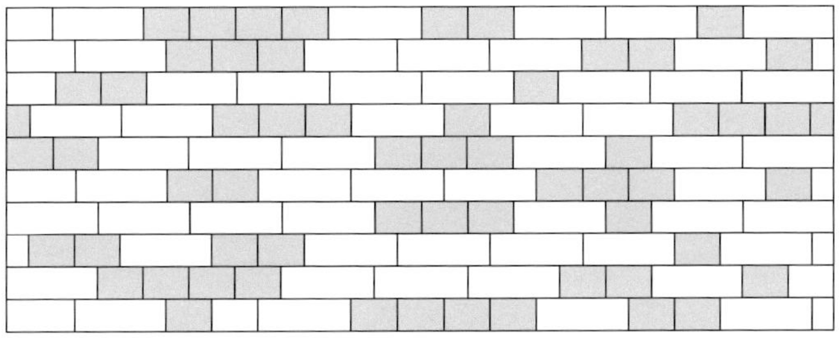

8

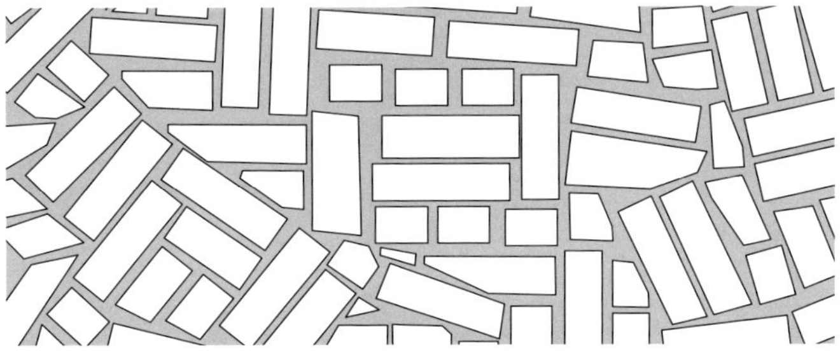

9

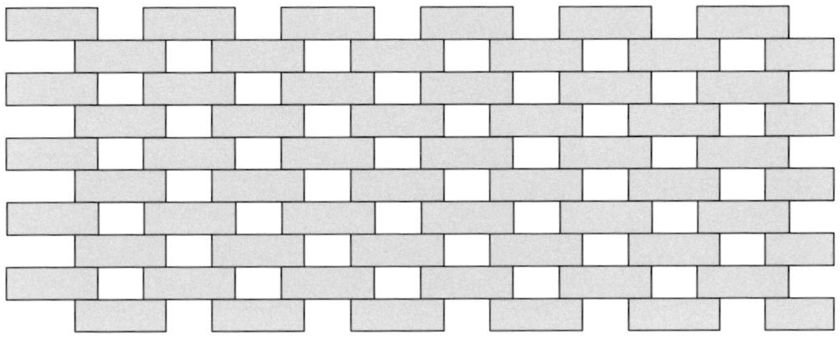

10

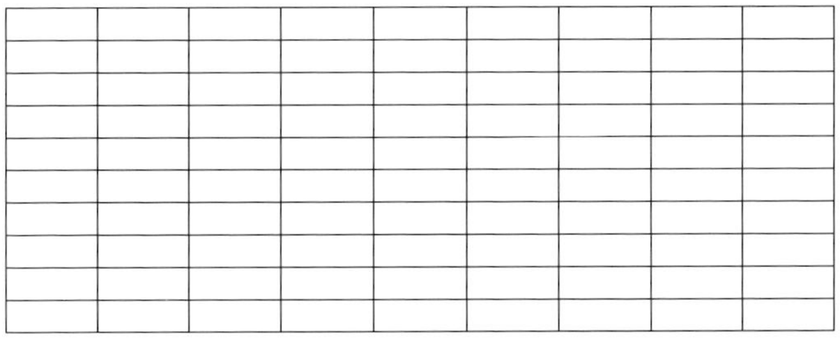

11

A wild bond is where the bricks are laid in a seemingly random formation, which delivers a cobbled, authentic appearance.

Honeycomb-style bonds

Various countries have local adaptations of the honeycomb-style bond, a decorative pattern that creates symmetrical voids within the brickwork. Claustra or Brazilian bond is a prime example of this honeycomb style. These bonds are used selectively in building facades to create special visual effects, allowing structures or light to be glimpsed through the gaps. They offer a plethora of creative opportunities by stacking facing bricks to generate open spaces, resulting in a translucent effect.

Stack bond or block bond

Stack bond, also referred to as Block bond, employs stretchers laid in vertical parallel rows without overlapping units. This pattern accentuates the vertical aspect of masonry. Note that fundamental brickwork principles discourage having less than a quarter-brick lap, making no-lap construction a skillful task.

Stack bond serves mainly aesthetic purposes or for infill panels, where structural strength comes from other sources. Even in such cases, extra bed reinforcement is recommended. In larger wall areas and load-bearing structures, steel reinforcement placed in horizontal mortar joints strengthens the wall. Precisely matched masonry units maintain vertical alignment. It is visually appealing and popular in modern designs, sometimes using contrasting white and black bricks. While widely used internationally, Stack bond is also gaining popularity in the United States.

Stack bond displays rows of stretchers, aligning each stretcher centrally with the one below. Vertical joints traverse the wall, ensuring continuous horizontal edge and vertical head joints. However, this type of bond lacks the inherent strength of most other bonds as the wall can crack vertically along the joint lines.

Facing bond

The Facing bond is employed in thick walls that incorporate facing and backing constructed using bricks of differing thicknesses. This bond involves alternating heading and stretching courses, with one heading course following several stretching courses. Since the number of joints varies between the backing and facing layers, load distribution is uneven, occasionally causing unequal settling.

Though challenging with pavers, [basket weave] pattern is feasible with standard bricks.

Basket weave bonds

Basket weave bond is commonly used in brick paving, sometimes with minor adjustments for walls. The width of the bricks, taking into account the joint, is slightly less than half the length. Though challenging with pavers, this pattern is feasible with standard bricks. While primarily used in flooring, it is occasionally employed in wall construction.

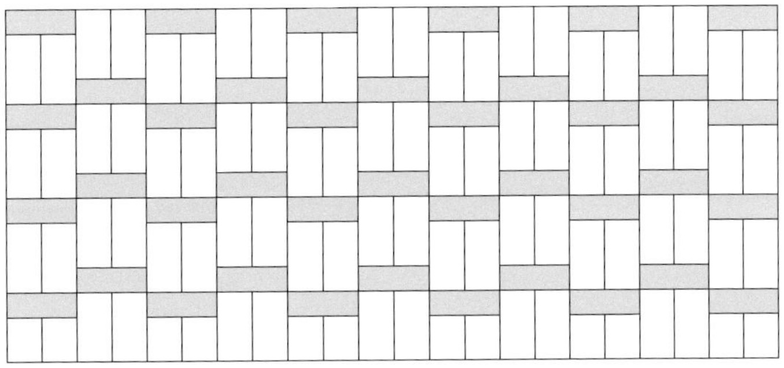

1 Single basket weave bond

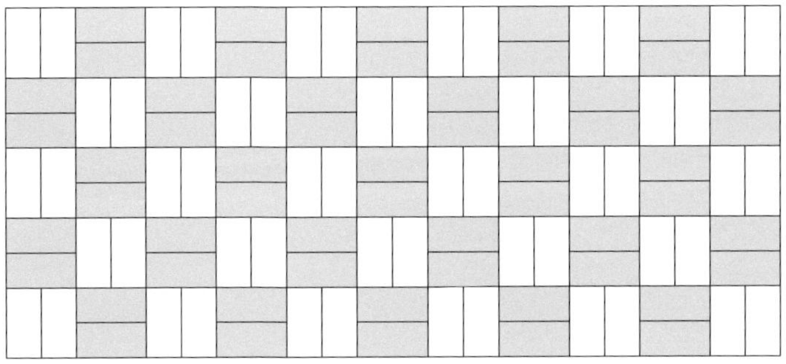

2 Double basket weave bond

Single basket weave bond This bond comprises pairs of sailors side by side, capped with a shiner (a brick laid on the long narrow side with the broad face of the brick exposed) alternating with pairs of sailors sat atop a shiner. Successive rows align with those above.

Double basket weave bond Rows of double basket weave bond involve pairs of shiners above one another, alternating with pairs of sailors side by side. The next row is offset, placing the shiners below the sailors in the preceding row. This arrangement yields bricks in pairs forming a square grid, with the joint between each pair perpendicular to the surrounding four pairs.

Tudor Revival bond The Tudor Revival bond is a variation of the basket weave pattern often found in UK Tudor Revival homes.

Raking bonds: Herringbone and Diagonal bonds

Walls that exceed a thickness of two bricks can experience a reduction in longitudinal strength due to the headers used within the wall's interior, which aim to enhance transverse strength. This issue is addressed by employing a raking bond, where 'rake' signifies an inclination. In this bonding technique, bricks are laid at an angle to the wall's surface.

There are two main variations of raking bonds: the Herringbone bond and the Diagonal bond.

Herringbone bond Herringbone bonds are commonly referred to as 'raking' bonds. They are typically used in conjunction with the English bond, as fitting diagonal bricks into Flemish facing would require extensive cutting. The Herringbone pattern involves positioning a sailor brick adjacent to a shiner brick to form an 'L' shape, and this combination is then repeated. This bond is most suitable for particularly thick walls, generally no less than four bricks in thickness. The pattern created by the raking bond is sometimes employed for brick laying on floors as well.

There are two variations of the Herringbone bond based on the angle of inclination:

- 90-degree Herringbone bond
- 45-degree Herringbone bond, also known as the Zig Zag bond. This variant features bricks laid in a zigzag pattern and is commonly used in brick-paved flooring.

Diagonal bond The Diagonal bond, also known as Diapering, Flemish Diagonal bond, or Diamond work, is utilized for walls that are 2 to 4 bricks thick. In this bond, face bricks are initially laid, followed by diagonal brick placement. The angle of inclination for the bricks should be adjusted to ensure they can be fitted without the need for cutting. When brickwork forms a diamond pattern, it is termed diapering. The Flemish diagonal bond encompasses a complex arrangement of stretcher courses alternating with courses of one or two stretchers positioned between headers. Over the span of ten courses, a diamond-shaped pattern emerges.

Brickwork built around square fractional-sized bricks

Pinwheel bond Pinwheel bond is made of four bricks surrounding a square half-brick, repeated in a square grid.

Della Robbia bond A pattern made of four bricks surrounding a square brick, one-quarter the size of a half-brick. It is designed to resemble woven cloth. Another, similar pattern is called the interlacing bond.

3

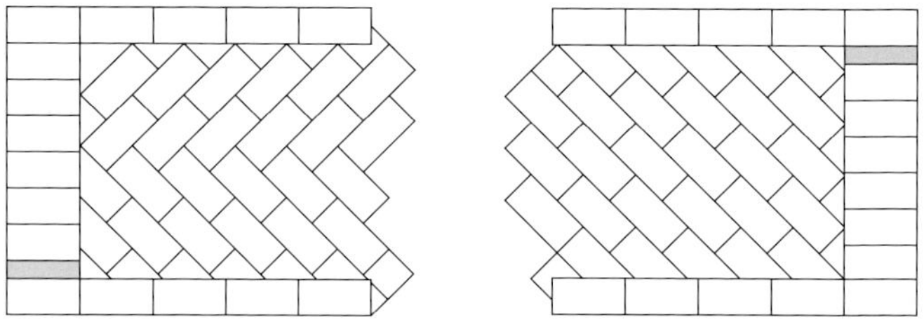

4

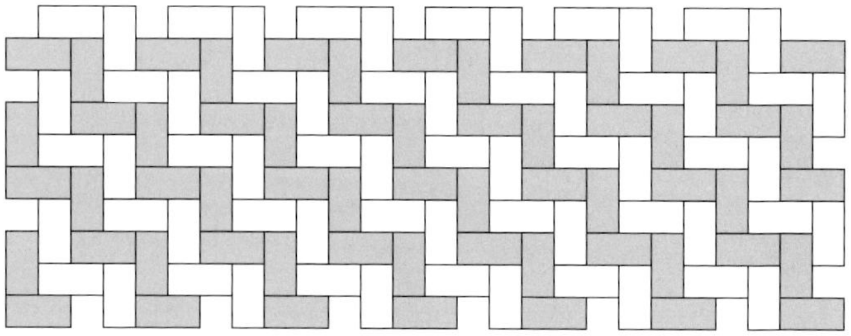

5

6

7

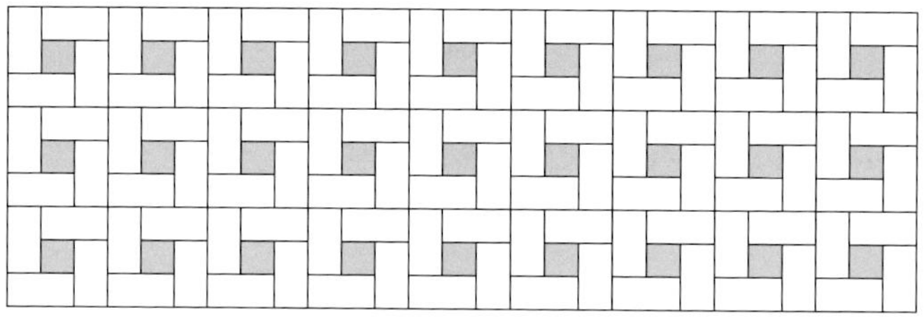

8

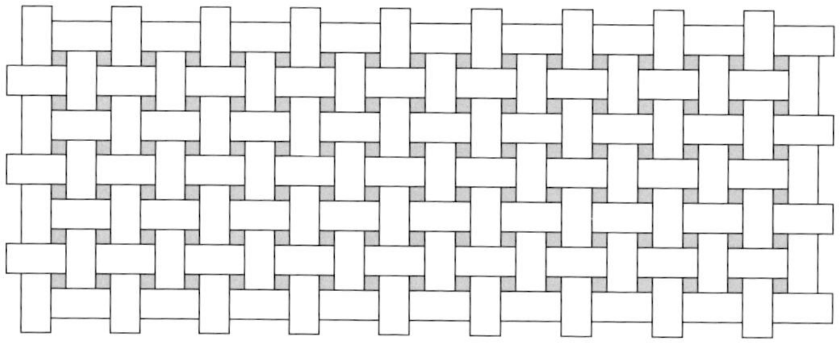

9

3 Tudor Revival bond
4 Raking bond - Herring bone bond, Diagonal bond
5 Herringbone 90° weave
6 Herringbone 45° weave or zig-zag bond
7 Diagonal bond
8 Pinwheel bond
9 Della Robbia bond

All images Detail of brickwork, Tabriz, East Azerbaijan Province, Iran. Mohammad Mohammadpour on Unsplash.

Rubble work

Rubble work is a particular form of construction more suited to stone walls that can be found in heritage buildings in the UK, parts of North America and continental Europe.

A similar result can be achieved by mixing stone and brick which creates a unique effect, but which requires much skills on the part of the brick mason or block layer.

Rubble work includes:

- Random rubble (uncoursed and built to courses)
- Squared rubble (uncoursed, built to courses and regular coursed)
- Miscellaneous (polygonal walling, flint walling, Lake District masonry)

Images 1, 2 & 3: McKay, W.B., Building Construction. William Barr McKay was born in Consett, County Durham in the UK and started his career in architectural training before 1914. He also qualified as a structural engineer, becoming a member of the Institute of Structural Engineers in 1927, and he served on the institute's council. McKay died in 1958.

For further information refer to the following publications (Vols 1, 2, 3) which are in the Public Domain: https://archive.org/details/W.B.McKayVol11945/page/n47/mode/2up

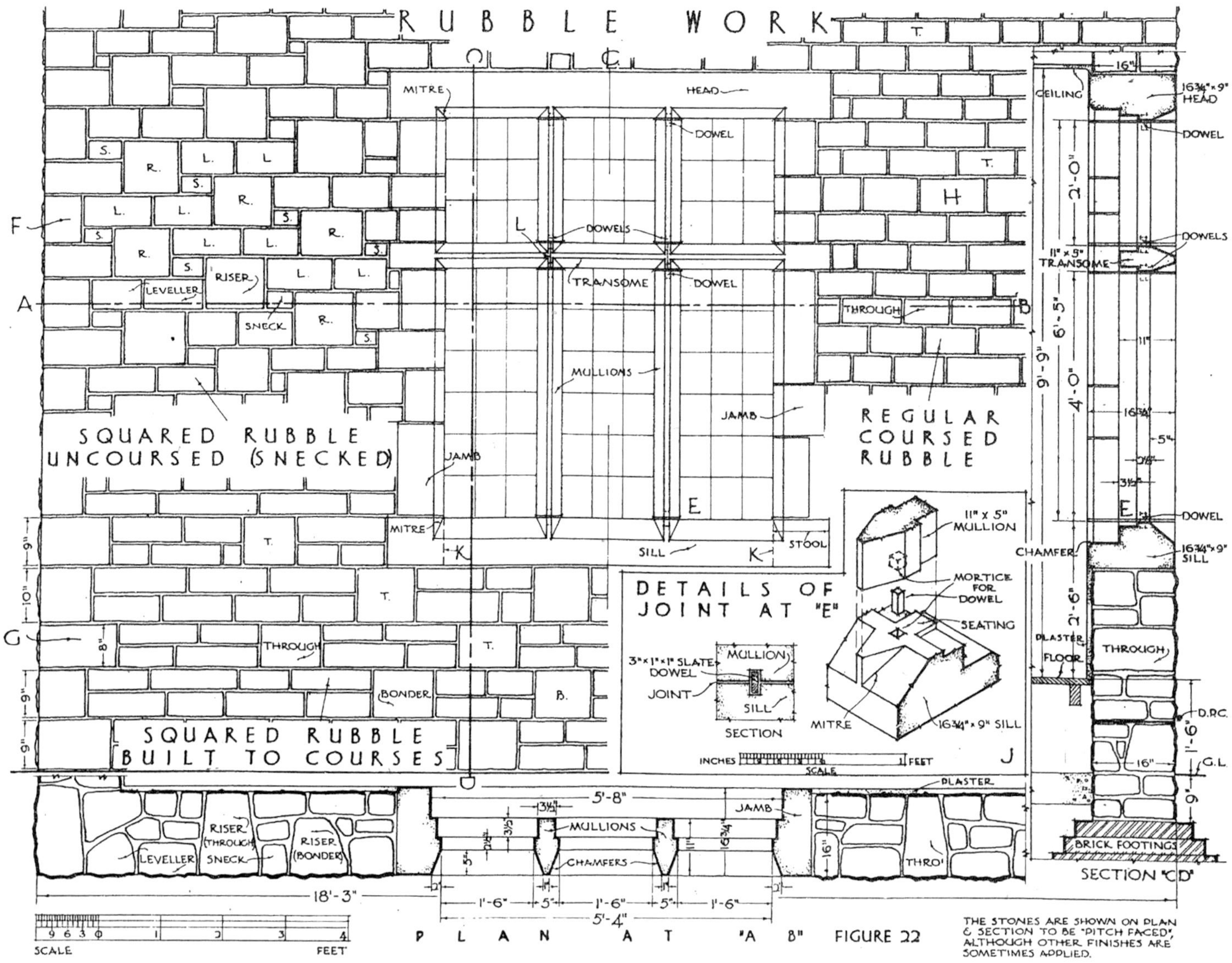

1

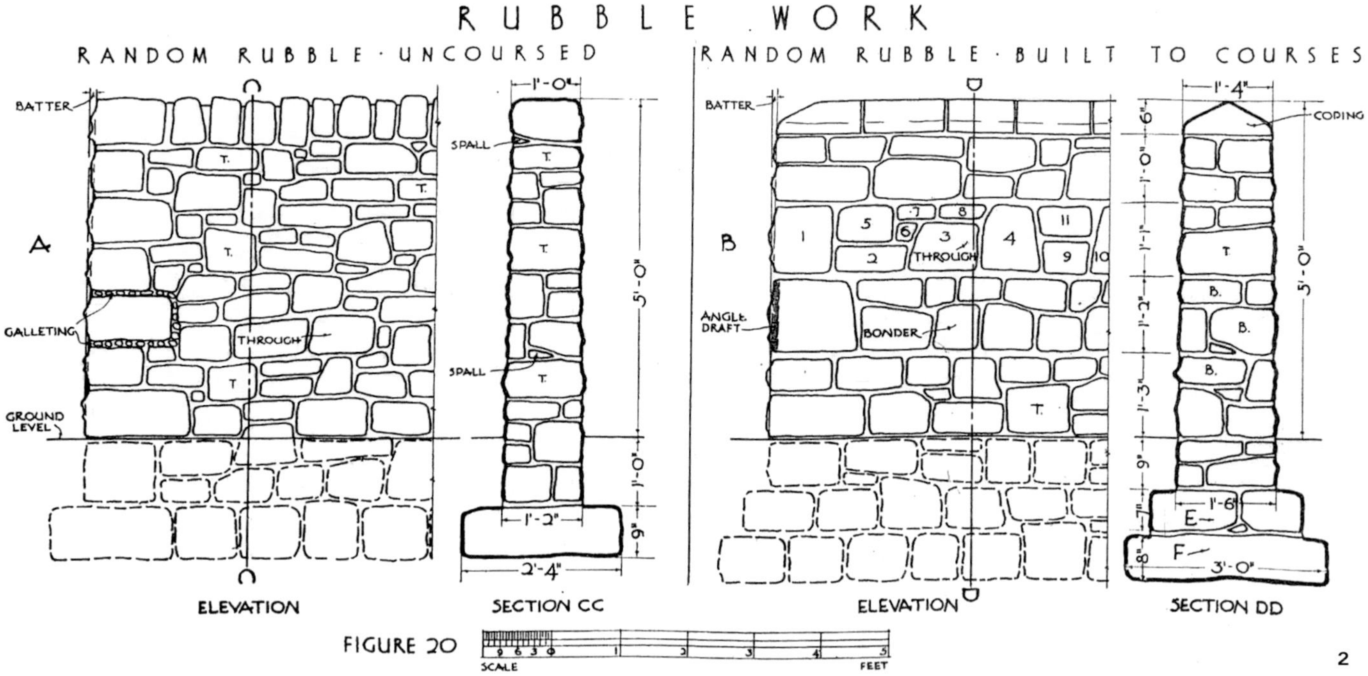
RUBBLE WORK
RANDOM RUBBLE · UNCOURSED
RANDOM RUBBLE · BUILT TO COURSES
BATTER
SPALL
GALLETING
THROUGH
GROUND LEVEL
A
B
T.
ANGLE DRAFT
BONDER
CODING
1'-0"
1'-2"
2'-4"
1'-4"
1'-6"
3'-0"
5'-0"
ELEVATION
SECTION CC
SECTION DD
FIGURE 20
SCALE
FEET

2

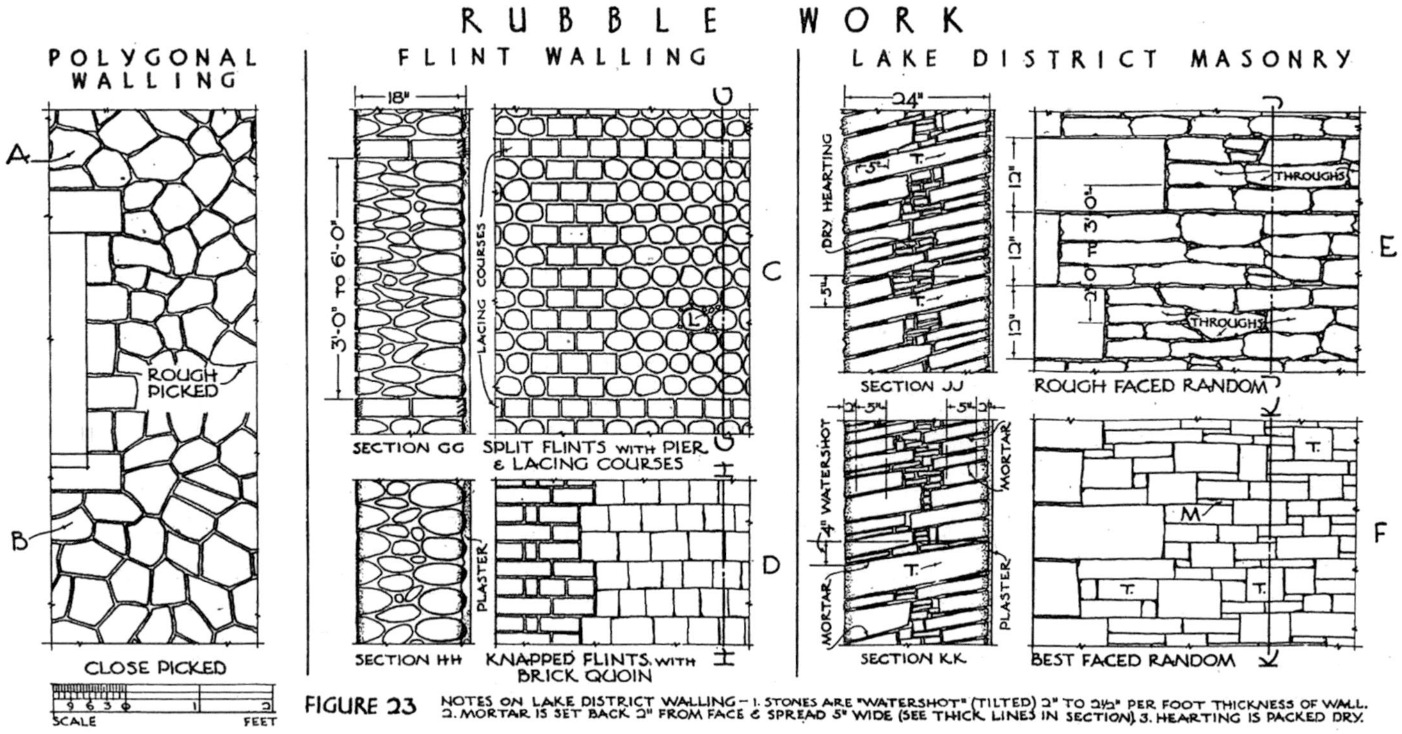
RUBBLE WORK
POLYGONAL WALLING
FLINT WALLING
LAKE DISTRICT MASONRY
A
B
C
D
E
F
ROUGH PICKED
CLOSE PICKED
LACING COURSES
3'-0" TO 6'-0"
SECTION GG
SPLIT FLINTS WITH PIER & LACING COURSES
PLASTER
SECTION HH
KNAPPED FLINTS WITH BRICK QUOIN
DRY HEARTING
24"
SECTION JJ
THROUGHS
ROUGH FACED RANDOM
WATERSHOT
MORTAR
SECTION K.K.
BEST FACED RANDOM
SCALE
FEET
FIGURE 23
NOTES ON LAKE DISTRICT WALLING – 1. STONES ARE "WATERSHOT" (TILTED) 2" TO 2½" PER FOOT THICKNESS OF WALL. 2. MORTAR IS SET BACK 2" FROM FACE & SPREAD 5" WIDE (SEE THICK LINES IN SECTION). 3. HEARTING IS PACKED DRY.

3

Quoins and cross walls in masonry construction

Quoins, substantial rectangular blocks made of masonry or brick, play a vital role in wall construction. These blocks are strategically placed at the corners of a wall, serving as both load-bearing elements for added strength and protection against weather conditions. Beyond their functional purpose, quoins are also used to elevate the visual appeal of a building by imparting intricate detailing to its outer corners.

Typically, quoins are arranged in a toothed pattern, with alternating blocks projecting and receding from the corners. When used for decorative purposes, a variety of materials like timber, cement render, and stucco can be employed. These materials are often colored to complement the overall brickwork or rendering of the building. Several types of quoins exist, each serving specific purposes:

- Return quoin: The standard type inserted into wall corners, with one long side aligned over one short side.
- Reveal quoin: Commonly used in door and window reveals, integrated as a standard block.
- Chamfered quoin: Featuring chamfered edges on the top, bottom, and outside edges for aesthetic enhancement. This design can be incorporated into reveal, return, or cladding-type quoins.
- Cladding quoin: Today, cladding quoin is attached to the wall using a two-part resin after corner completion, typically with a thickness of 50mm. Traditionally, mortar was used to affix the cladding quoin. When rendering is planned, the quoins are affixed first, followed by bringing the render up to the chamfer.

Cross walls, English bond When cross walls meet the main walls obliquely, a slight modification is required to the above method to secure good bonding. Where one course of the cross wall penetrates the main wall, the headers must be as large as possible. Pieces of brick cut to various shapes as required must be inserted in place of the queen-closers in right-angle junctions. The pieces should be as large as possible for strength (and also aesthetics), but not large enough to reduce the amount by which the cross wall overlaps the main wall.

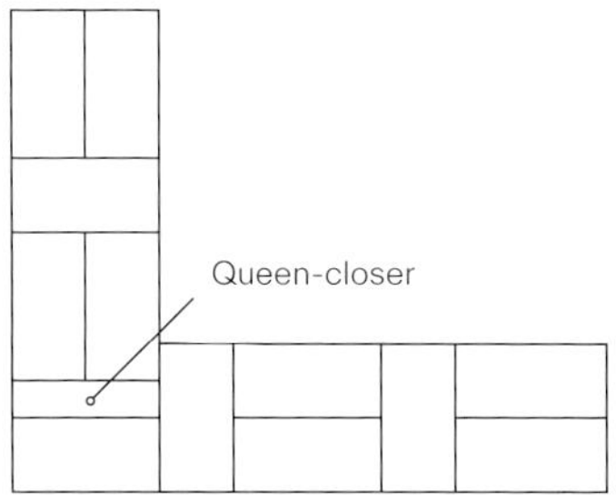

Quoin - plan of first course

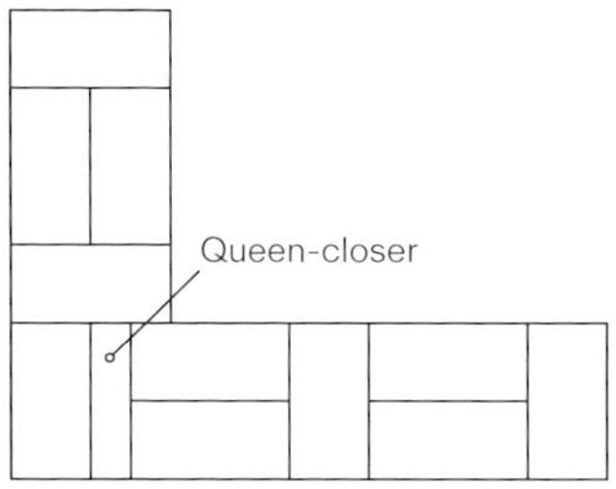

Quoin - plan of second course

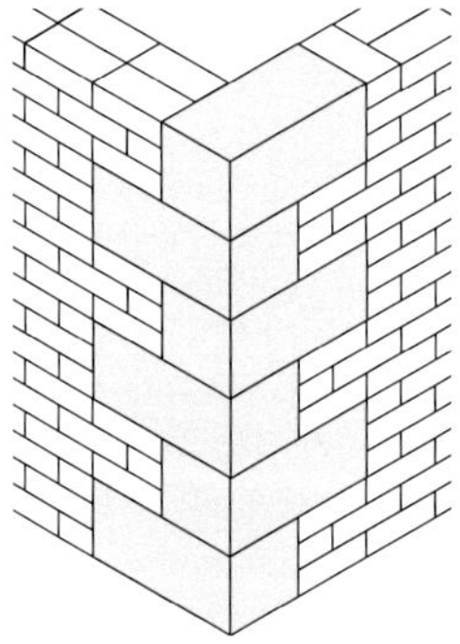

Return quoin

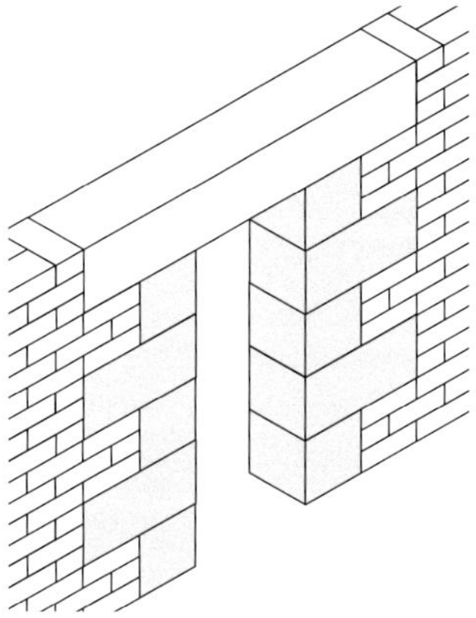

Reveal quoin

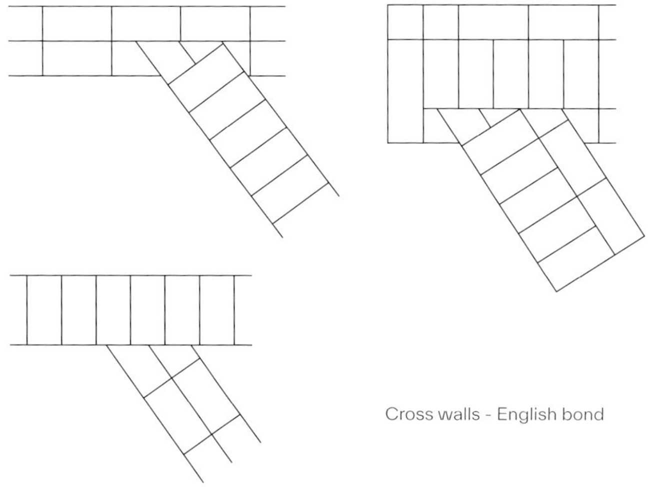

Cross walls - English bond

Cavity wall construction

Cavity walls are consistently constructed using stretcher bond. Cavity ties, stipulated by architects or designers, are indispensable for imparting strength to the wall. These ties, available in various patterns and materials, feature a kink in their center that must always point downward to allow moisture to drip off and not traverse to the dry inner skin. The bricklayer embeds these ties in the first skin while progressing with the work, extending to the second wall as it takes shape, adhering to a staggered pattern.

The damp-proof course (DPC) varies in materials and is usually positioned a couple of courses above ground level. Internally, the DPC is situated one course higher than the exterior, facilitating moisture escape through weep holes. This prevents rising damp as bricks are porous and absorb water through capillary action. Weep holes (also referred to as perpends) found every third brick or so remain mortar free.

Maintaining a gap of 50mm or 2", the cavity must be clean, devoid of mortar dags bridging the gap between brick skins so that water does not travel to the inner dry skin. The standard procedure involves erecting all walls to DPC height and filling the cavities up to that level with concrete.

Walls: Wythes and Structure A wythe, a continuous vertical section of a wall comprising a single masonry unit, holds a significant place in construction. This section, also referred to as a tier, can be independent or interlocked with adjacent wythes. A non-structural single wythe of brick is termed a masonry veneer. Multi-wythe masonry walls might consist of a single masonry unit type layered for increased thickness and strength, or a combination of different units selected for their function, such as cost-effective concrete blocks for structural purposes and more visually appealing bricks for appearance.

A non-structural single wythe of brick is termed a masonry veneer.

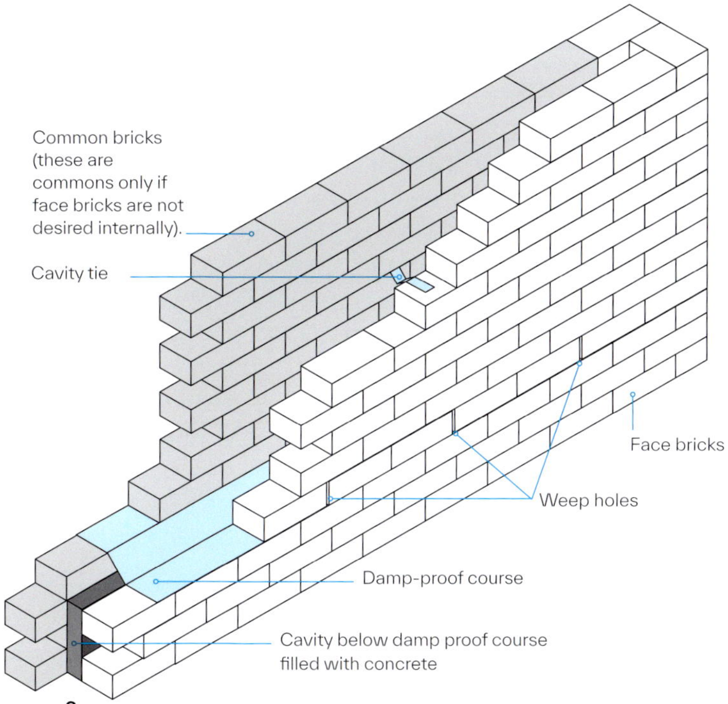

3

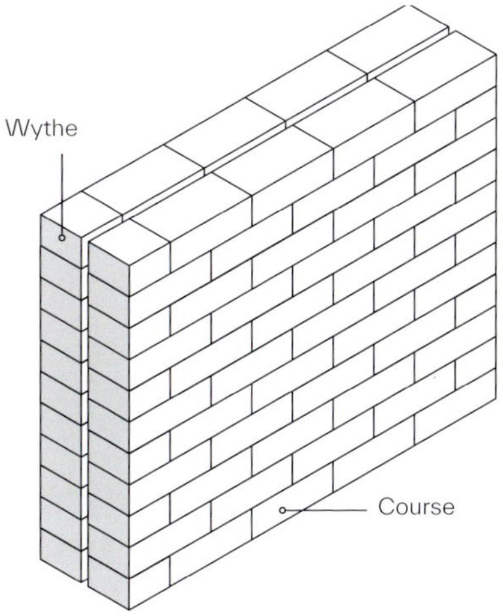

4

Conclusion

Bricks hold immense significance in contemporary construction due to their enduring attributes and eco-friendly nature. They represent a sustainable building material, often made from locally-sourced clay or shale. Compared to alternative materials like concrete, bricks require lower energy inputs during production, reducing their environmental footprint. Their durability ensures that structures built with bricks can withstand the test of time, contributing to long-lasting, sustainable infrastructure.

Furthermore, bricks contribute to structural resilience. They offer excellent thermal mass properties, regulating indoor temperatures and reducing the need for excessive heating or cooling. This efficiency translates into energy savings and lower utility costs for occupants.

Beyond their environmental benefits, bricks offer unmatched versatility in design. They can be arranged in various patterns, creating unique and aesthetically pleasing structures. Their adaptability extends to different architectural styles and contexts, seamlessly blending with both historic and contemporary surroundings.

One crucial aspect of brick construction is the selection of appropriate brick bonds. Brick bonds dictate the arrangement and interlocking pattern of bricks in a wall, influencing its stability and appearance. The bond's layout influences a wall's load-bearing capacity, distribution of stress, and resistance to lateral forces. Ultimately, the choice of brick bond is integral to a building's integrity and aesthetics. It showcases the craftsmanship and attention to detail of a construction project, emphasizing the enduring importance of bricks and their bonds in modern architecture.

In a world increasingly focused on sustainable, resilient, and aesthetically pleasing construction, bricks continue to play a pivotal role, embodying a perfect synergy of tradition and innovation. Their importance today lies not only in their practical utility but also in their ability to meet the evolving demands of a forward-thinking society.

Sources

The author wishes to acknowledge the use of information from the following sources.

ACME Brick, https://brick.com/special-brick-shapes

Batchelder & Collins, https://757brick.com/a-guide-to-brick-shapes-and-sizes/

BigRentz, https://www.bigrentz.com/blog/brick-dimensions

Bill Bradley, http://www.builderbill-diy-help.com

Brick Architecture, https://brickarchitecture.com/about-brick/why-brick/brickwork-bonds

Civil Engineering, https://civiltoday.com/civil-engineering-materials/brick/271-brick-size-standard-dimensions

Nikolas Davies & Erkki Jokiniemi, *Dictionary of Architecture & Building Construction*, Elsevier, London, 2008.

Designing Buildings, https://www.designingbuildings.co.uk/wiki/Brick_sizes

General Shale, https://generalshale.com/custom-brick-shapes/

Gharpedia, https://gharpedia.com/blog/various-shapes-of-brick-used-in-construction/#:~:text=The%20shapes%20of%20brick%20may,other%20shape%20in%20cross%2Dsection.

Glen-Gery, https://glengery.com/

Homestratosphere, https://www.homestratosphere.com/types-of-bricks/

Old Carolina® Brick Company, https://www.handmadebrick.com/dir.cfm/BRICK_OPTIONS/SHAPES/

McKay, W.B., Building Construction, https://archive.org/details/W.B.McKayVol11945/page/n47/mode/2up

NCMA, https://ncma.org/resource/typical-sizes-and-shapes-of-concrete-masonry-units/

Triangle Brick, https://www.trianglebrick.com/resources/shapes-guide

US Brick, https://usbrick.com/shapes-and-sizes/

Vandersanden Group, https://www.vandersanden.com/en-uk/bricklaying-pattern

Wienerberger

Wikipedia

The choice of brick bond showcases the craftsmanship of a construction project, emphasizing the enduring importance of bricks and their bonds in modern architecture.

Above Tate Modern by Herzog & de Meuron, London, United Kingdom. Image: Samuel Regan-Asante on Unsplash.

Right Billerica Memorial High School, Perkins&Will. Image: James Ewing.

Masonry arches
Ron Ringer

Author's note on sources: This feature on arches is indicative of just some of the types of arches used in construction today and is not intended to be comprehensive. The author acknowledges the selective use of material that first appeared in Nikolas Davies & Erkki Jokiniemi, Dictionary of Architecture & Building Construction, *Elsevier, London, 2008. A variety of other sources have also been used.*

Arches are compressive structures in that they possess no tensile stresses, but are self-supporting and stabilized by the force of gravity acting on their weight to hold them in compression. This makes them very stable and efficient, capable of larger spans and able to support greater loads than horizontal beams. Arches may be synonymous with vaults; however, a vault may be distinguished as a continuous arch forming a roof.

Brick arches, in particular, have an ancient history and first appeared as early as the second millennium BC in the mud brick architecture of Mesopotamia. Systematic use of arches started with the ancient Romans, who were the first to apply the technique to a wide range of structures built using brick and stone.

Gothic arches convey something grand and stately. This style of architecture flourished in Europe from the mid-twelfth century to the sixteenth century. Masonry buildings in the Gothic style (whether in the early middle ages or the period of Gothic Revival in the later decades of the eighteenth and nineteenth century in Great Britain) were cavernous spaces with their expanse of walls broken up by overlaid tracery.

Brick arches are elegant and can be built in several different shapes and styles to give windows and doorways a graceful, finished look. Arches are visually uplifting, welcoming and can be simple and understated or bold and dramatic. Keys

are a classic feature of some types of arches and can help to create a sense of tradition and substance. The smooth surfaces contrast beautifully with the textures of surrounding brick. Brick arches, however, are expensive to build in terms of labor costs.

Louis Kahn, one of the most influential architects of the twentieth century, was celebrated for his use of brick in design and construction. The child of poverty-stricken immigrants from Estonia who first settled in Philadelphia, Kahn's birthname was Leiser-Itze Schmulowsky, which he later changed. By an unusual—and sad—twist of fate he died in 1974 at the age of 73 of a heart attack in a bathroom at New York's Penn Station. Kahn's body went unclaimed for four days. Strangely, the name on his passport had been obliterated. Bankrupt at the time of his death, Kahn's small Philadelphia practice was $500,000 in debt.

Yet, Kahn's work was legendary. The Salk Institute in California, Yale University Art Gallery, and the National Assembly building of Bangladesh are just three examples of his exquisite work in design and construction. Kahn had a lifelong fascination with brick and its design possibilities, enjoining his students to consider: 'You say to brick, "What do you want, brick?" Brick says to you, "I like an arch." If you say to brick, "arches are expensive and I can use a concrete lintel over an opening. What do you think of that, brick?" Brick says: "I like an arch."'

Left Wells Cathedral, Wells, Somerset, England. Image: Annie Spratt on Unsplash.

Above Sultan Qaboos Grand Mosque in Muscat, Oman. Image: Efired, iStock by Getty images.

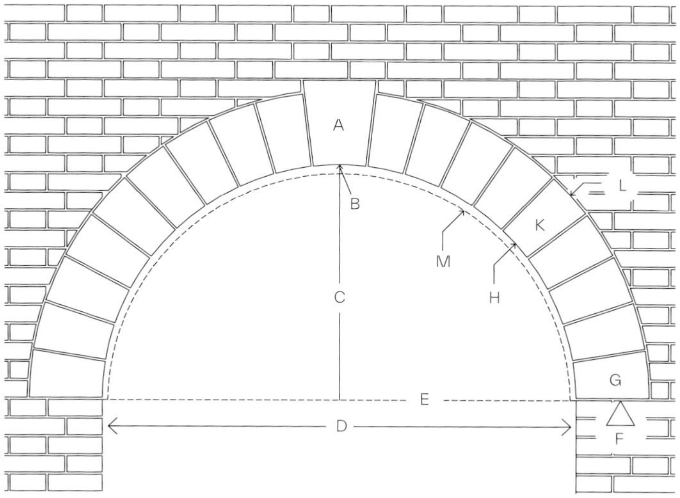

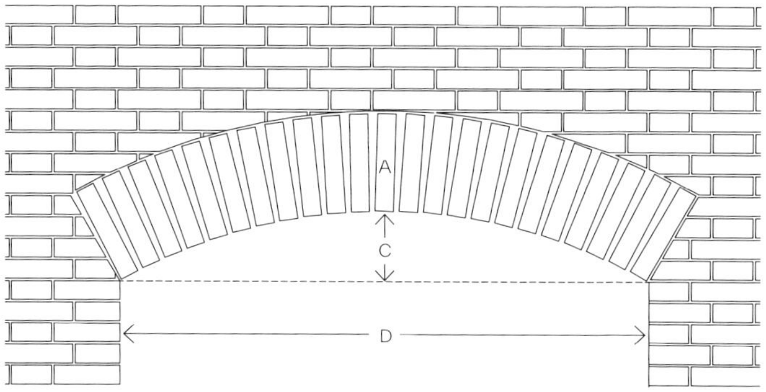

Arched and vaulted construction

An arch is a two-dimensionally curved beam construction for supporting loads between two points of support over an opening. The construction of traditional brick arches is dependent on the arrangement of the bricks over the opening. Wedge-shaped blocks (voussoirs) are set flank-to-flank, with the upper edge being wider than the lower edge. Downward pressure on the arch has the effect of forcing the voussoirs together instead of apart. The voussoir positioned in the center of the arch is known as the key. In ordinary brickwork there are two standard types of arches: the rough segmental arch and the gauged (or cut) semi-circular arch.

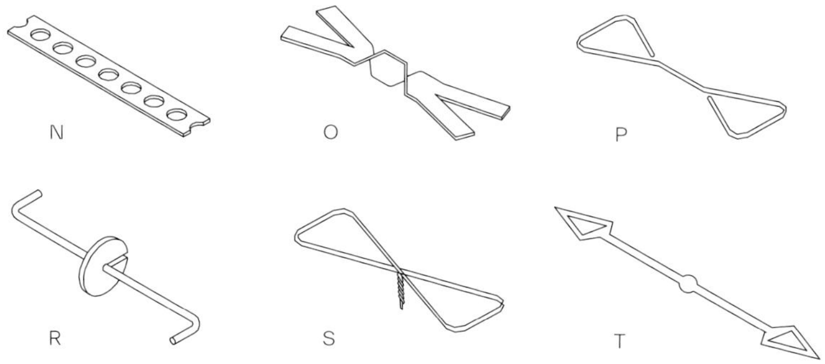

Composition of an arch

A keystone, key block
B crown
C rise, pitch
D span
E springing line
F abutment, impost
G springer, springing stone, skewback
H intrados, soffit
K voussoir
L extrados
M haunch

Enhancing stability – wall ties and anchor straps

A wall tie is a galvanized steel or twisted wire fixing that is laid at regular intervals into a masonry wall. It provides stability by connecting masonry work to a structural base, or linking inner and outer leaves of brickwork; called a cavity tie when used in cavity walls

N anchor strap
O fishtail tie
P double or twin triangle tie
R wire tie with insulation clip/retainer
S butterfly tie
T proprietary wall tie

Constructing an arch

Traditional masonry arches were constructed from wedge-shaped stones locked together by loading from above, although brick has been used for millennia to construct arches. To support a masonry arch during its construction a temporary curved frame of wood or other material is used.

1 centering for semi-circular arch construction
2 center, centering
3 shuttering, sheathing
4 strongback, spreader beam, yoke
5 strut, prop
6 centering for segmental and camber arch construction
7 turning piece

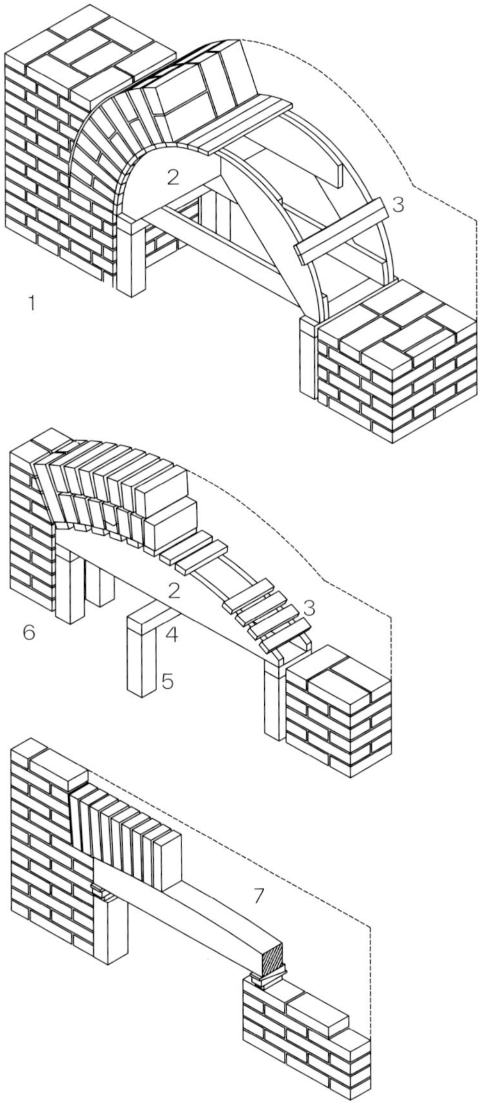

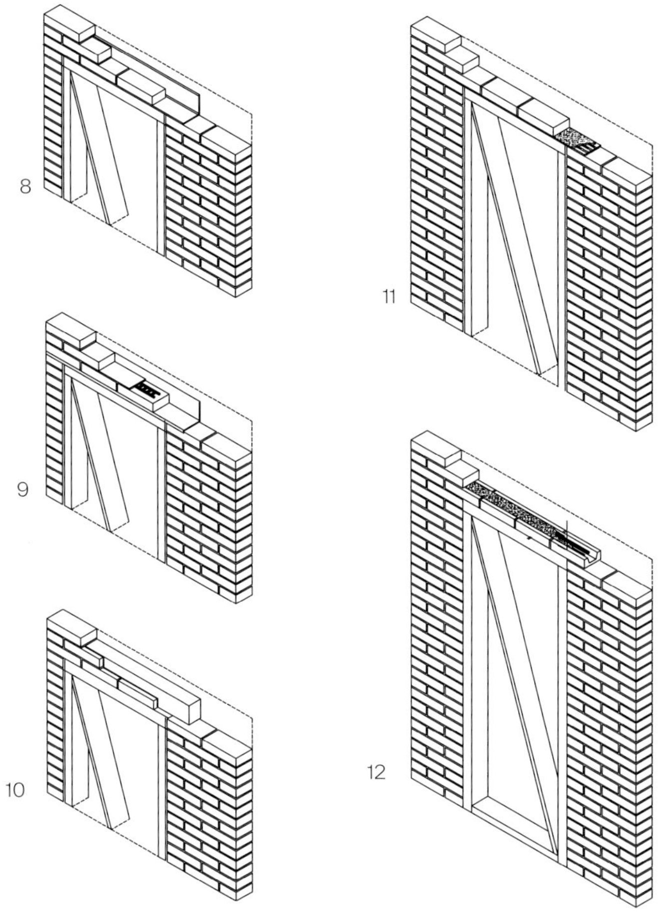

Lintels

A lintel is a beam above a window or door opening and is widely used in housing. The flat span is formed by vertical (soldier like) stretchers, often built with a support backing. For a detailed explanation and description of 'soldier' orientation of a brick, refer to the feature article in this section: Brick masonry bonds. One example of this type of brick arch construction technique is the soldier arch, which is not a true arch but rather a joggled lintel. Thus, soldier arches are not composed of elements that support each other while creating the form of a curve, which is why they are not technically arches. The flat arch has the long sides of the uncut bricks (also referred to as stretchers) set vertically. Early types of this sort of arch were reinforced by mild steel rods which passed through perforations in the bricks. Later types are supported by special projections on pre-cast concrete or steel lintels.

8 pressed steel lintel
9 proprietary reinforcing strap
10 pre-cast concrete lintel with brick slip facing
11 reinforced bed joint
12 reinforced brick lintel, brick beam

A rough chronology

The immediate predecessor of the True arch was the corbelled arch. Between the seventh and tenth centuries builders started adding more rooms, doors and windows to buildings. These were made by placing a horizontal beam across two vertical columns, a style of architecture called 'trabeated' or 'corbelled'. The projecting slabs used as corbels extended closer and closer to the opposite post while supporting the lintel from below.

The corbelled technique relied on building with large blocks of stone fitted into each other and believing that the weight of massive blocks gave strength to structures. This style made almost no use of binding materials such as mortar.

The following is a guide only and provides a rough chronological order to describe the development of the arch.

- Triangular arch
- Round or semicircular arch
- Segmental arch (less than a semicircle)
- Unequal round or rampant arch
- Lancet arch
- Equilateral arch
- Shouldered flat arch (see also jack arch)
- Trefoil or three-foiled cusped arch
- Horseshoe arch
- Three-centered arch
- Elliptical arch
- Inflexed arch
- Ogee arch
- Reverse ogee arch
- Four-centered or Tudor arch
- Parabolic arch

False arches

False arches such as triangular or corbelled arches do not utilize masonry voussoirs to provide support over an opening.

1. Triangular arch A triangular arch (also called a miter arch) is a simple arch with a short span that is often formed by two large diagonal stones that mutually support each other to span an opening.

2. Corbel arch, corbelled arch This structure resembles an arch in which successive courses project farther into a gap until they close. The corbel which gives it name to this type of arch is a masonry bracket projecting from the face of a building surface to provide support for an overhanging object or member.

3. Corbelled lintel The illustration shows a variation on the corbelled arch, except that the corbel underpins the lintel. See previous, Lintels, which illustrate door openings rather than window apertures.

Flat arches, Jack arches, Straight arches

These arches are so-called because they are flat on the bottom, which allows the arch to support the weight of the brick above it. All of these variations are straight along the springing line with a flat intrados (the lower line, underside, soffit or face of an arch). The masonry units are laid out as voussoirs angled to a center below the arch. See page 320 for examples of flat arches.

4. Rough brick arch This arch is made up of rectangular bricks that are neither cut nor tapered to voussoir shape; the required curvature is achieved by additional mortar in the joints.

5. Gauged arch This type of arch is constructed from soft, finely cut wedge-shaped bricks which have been shaped so that the joints radiate from a common center. The bricks are fashioned from a design template that is used to produce the correct shape. These special bricks (known as 'rubbers') are fired at a low temperature to make them soft enough to shape. They are then repeatedly rubbed and measured (or 'gauged') until they are the proper size. Due to the nature of this process, this type of arch can be quite costly and requires a significant degree of skill. In some instances, it may be possible to use specially made hard bricks in gauged arches.

6. Dutch arch, French arch, Welsh arch This arch is technically a flat arch in brick where most of the bricks slope outward from the middle of the arch (at the same angle on both sides of the centerline) and do not have radial joints.

The French arch uses standard brick widths and only the ends need to be cut. It was practical in the days when these arches were popular for the brickmakers to make purpose-made arch bricks with the ends cut to the skewback angle for these types of flat arches. No cutting or shaping is required and the span of the arch does not alter the shape of the bricks. In some cases with these French arches a single purpose-made stone key would be used.

7. Joggled arch, crossette This is where decorative embellishment is used, such as a molding around one corner of a door, window, or fireplace opening. The molding resembles a squared-off ear, and was especially popular during the latter half of the eighteenth century. Also called a dog's ear.

8. Pseudo three-centered arch An arch whose inner curved surface is struck from three centers, resulting in a shape approximating one-half an ellipse. Variation on a two-centered arch.

9. Jack arches

10. Pseudo four-centered arch

Left Alhambra, Calle Real de la Alhambra, Granada, Spain. Image: Izuddin Helmi Adnan on Unsplash.

Right Arc by Koichi Takada Architects, Sydney, NSW, Australia. Image: Mitchell Luo on Unsplash.

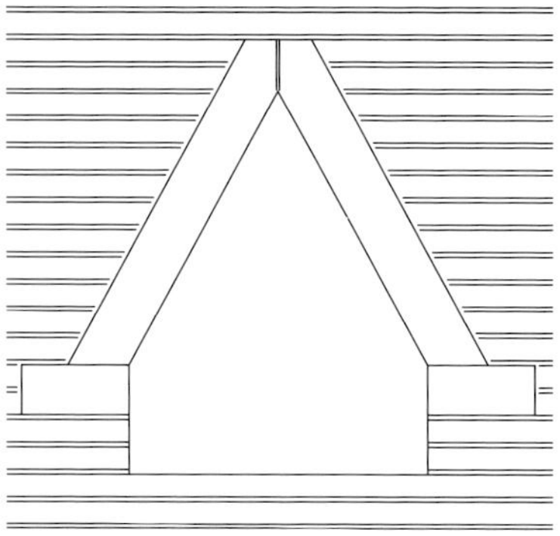

1

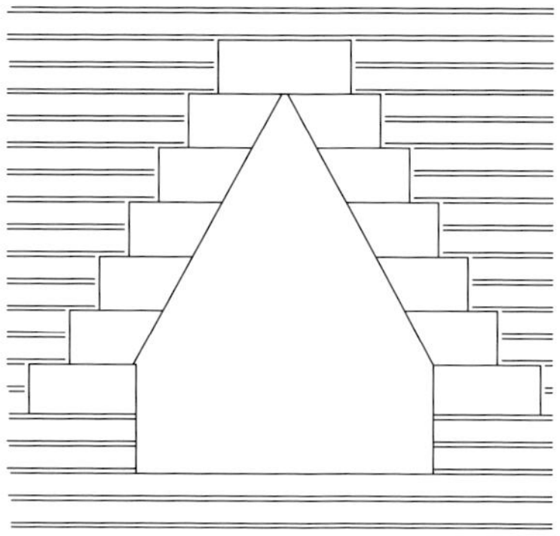

2

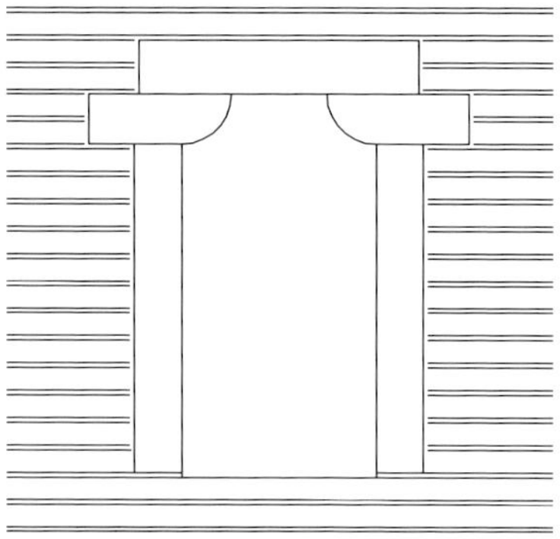

3

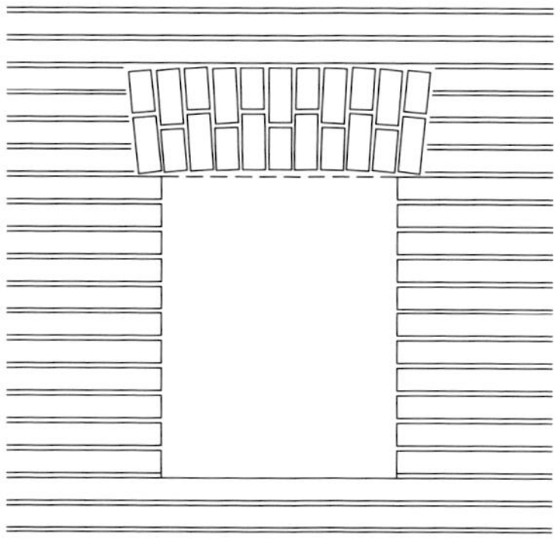

4

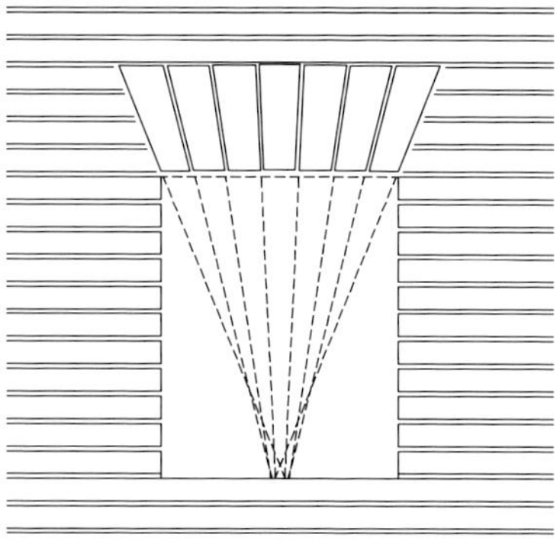

5

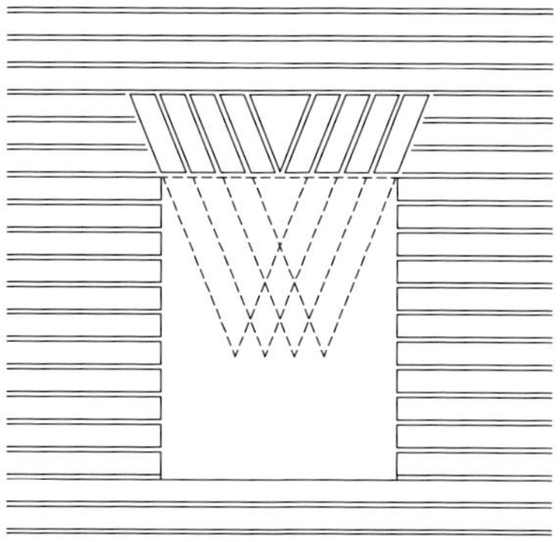

6

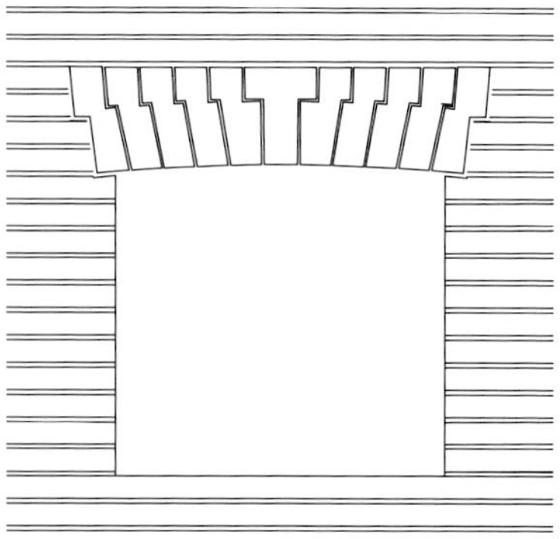

7

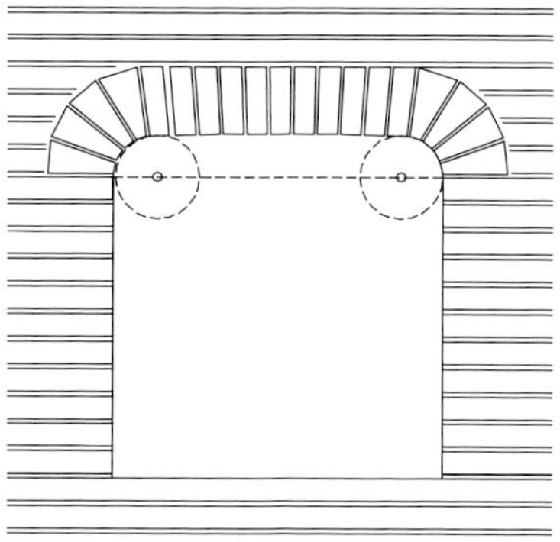

8

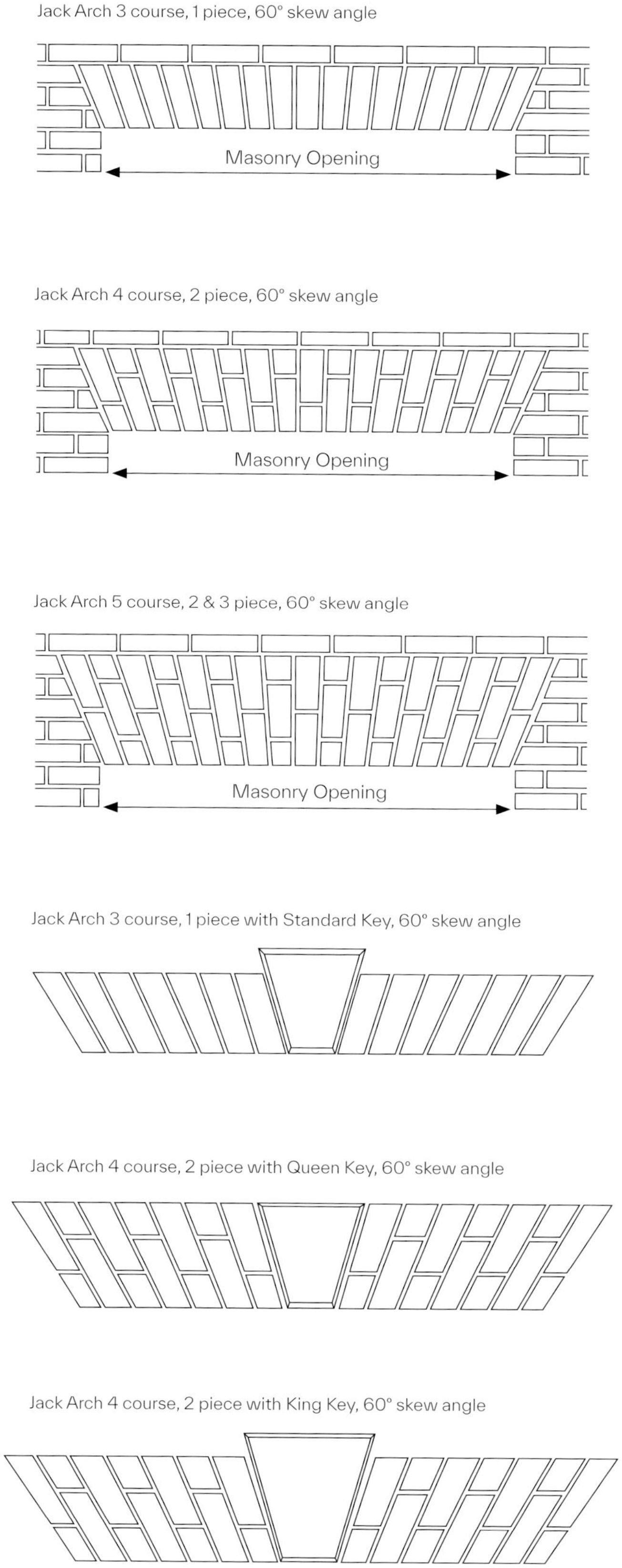
Jack Arch 3 course, 1 piece, 60° skew angle
Masonry Opening
Jack Arch 4 course, 2 piece, 60° skew angle
Masonry Opening
Jack Arch 5 course, 2 & 3 piece, 60° skew angle
Masonry Opening
Jack Arch 3 course, 1 piece with Standard Key, 60° skew angle
Jack Arch 4 course, 2 piece with Queen Key, 60° skew angle
Jack Arch 4 course, 2 piece with King Key, 60° skew angle

9

Italian arches

11. Venetian arch A Venetian arch is a pointed arch in which the intrados and extrados (the upper line of the voussoirs in an arch) are farther apart at the peak than at the springing line.

12. Florentine arch A semi-circular arch where the extrados (the upper or outer curve of an arch) is struck from a higher point than its intrados. Hence, the length of the voussoirs is longer nearer the top of the arch.

13. Bell arch A round arch supported on large corbels, giving rise to a bell-shaped appearance. Also referred to as a corbel impost block.

14. Relieving arch, Discharging arch A support built over, but not touching, a weak structural member, such as a wooden lintel, to carry the main load. Also known as relieving arch. This type of arch is designed to take the weight of masonry above it off another arch or lintel.

15. Strainer arches An arch placed between two walls to prevent them from leaning toward each other.

16. Roundel, circular arch, bullseye arch A roundel, sometimes known as a bullseye arch, is an ornamental circular window or oculus. This type of arch is a double arch with two or four keystones.

14

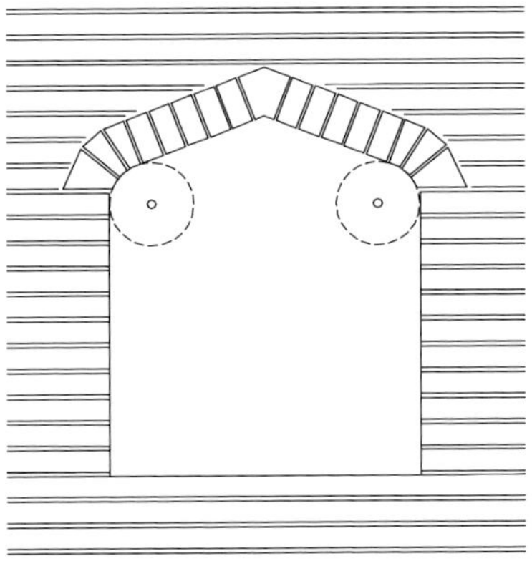

10

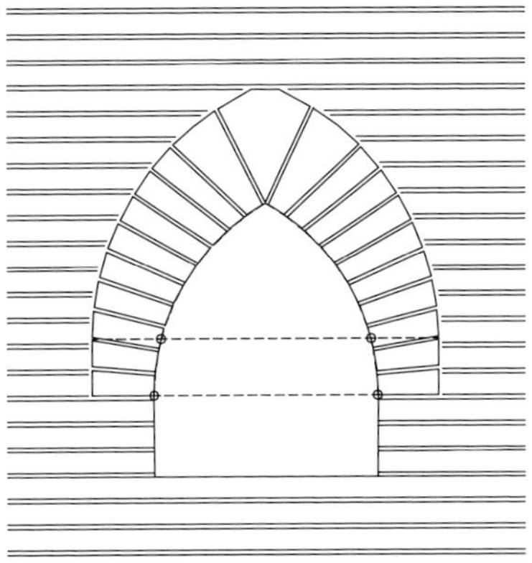

11

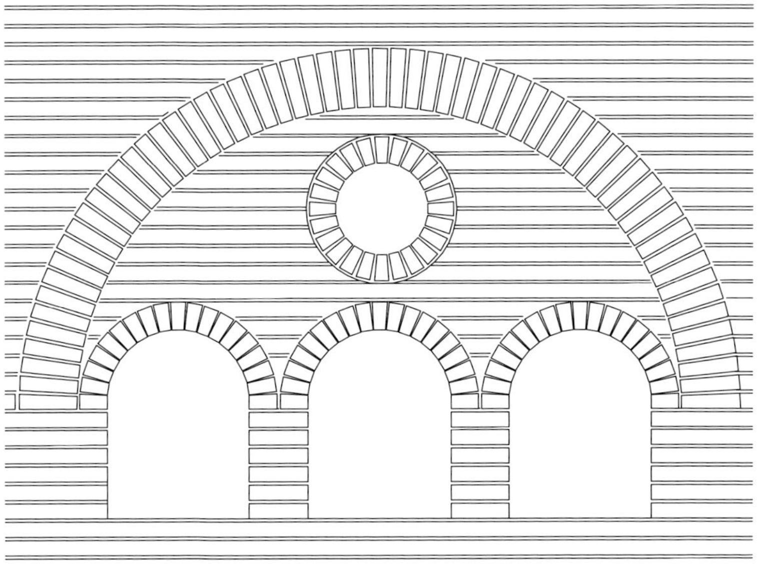

15

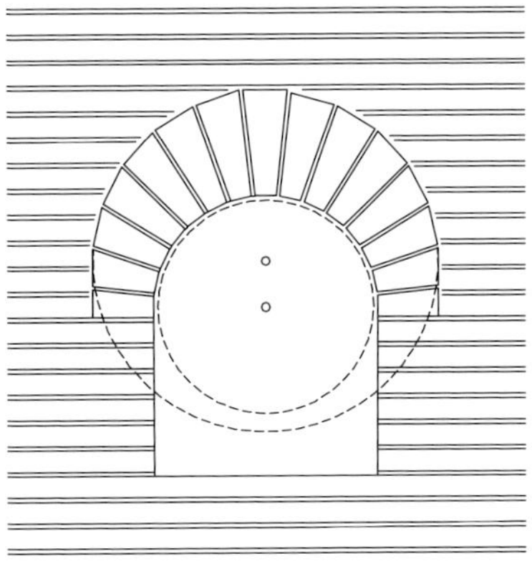

12

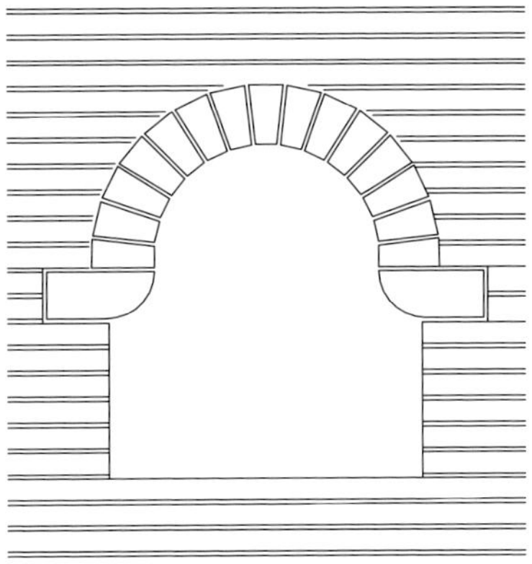

13

1 Triangular arch
2 Corbel arch, corbelled arch
3 Corbelled lintel
4 Rough brick arch
5 Gauged arch
6 Dutch arch, French arch, Welsh arch
7 Joggled arch, crossette
8 Pseudo three-centered arch
9 Jack arches
10 Pseudo four-centered arch
11 Venetian arch
12 Florentine arch
13 Bell arch
14 Relieving arch, Discharging arch
15 Strainer arches

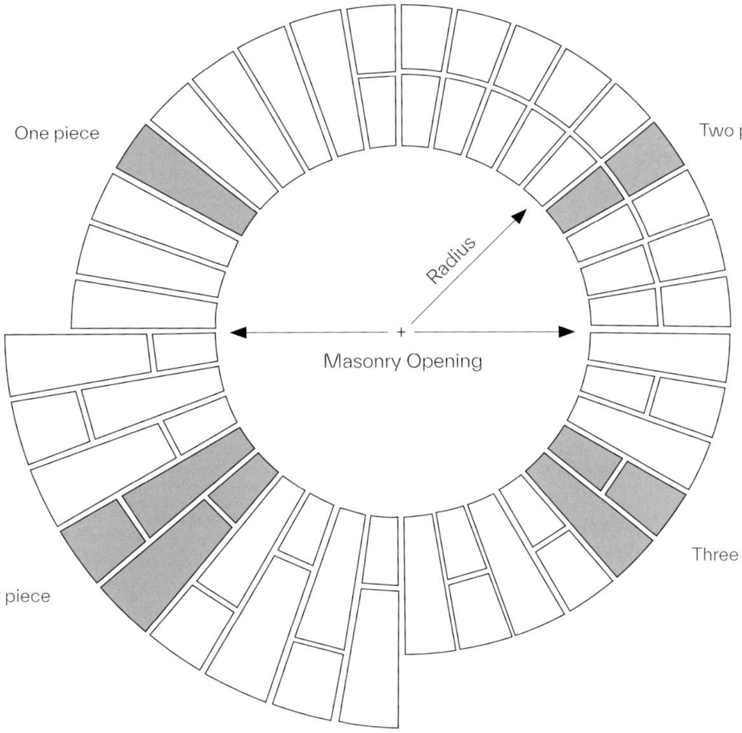

16

Decorative arches

17-22. Foliated arches Foliated arches have been carved or rendered with decorative leaf motifs, or foils radiating outwards from a point. They are found especially in medieval architecture. Stonework, rather than brickwork, characterizes these types of decorative arches.

16 Roundel, circular arch, bullseye arch
17 Pointed trifoliated arch
18 Pointed cinque foliated arch
19 Pointed multifoliate arch
20 Round cinque foliated arch
21 Round trifoliated arch
22 Round multifoliate arch

17

18

19

20

21

22

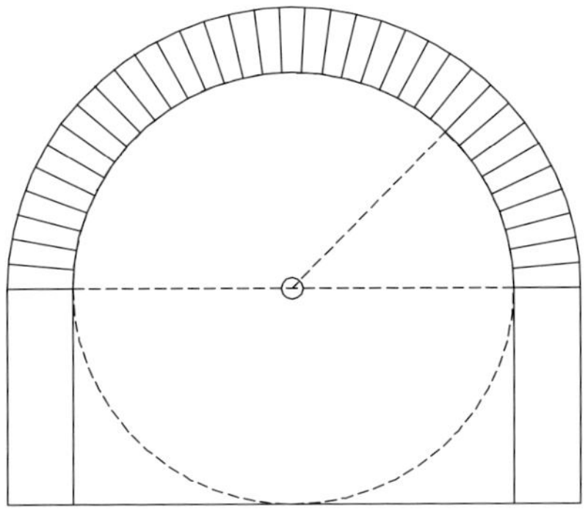

1a

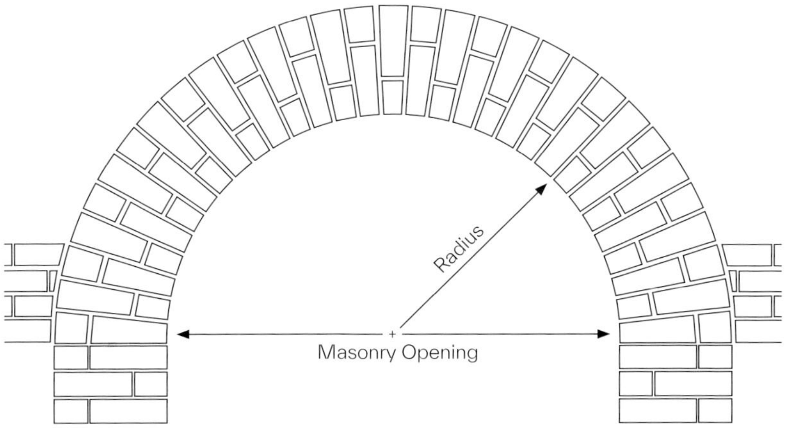

1b

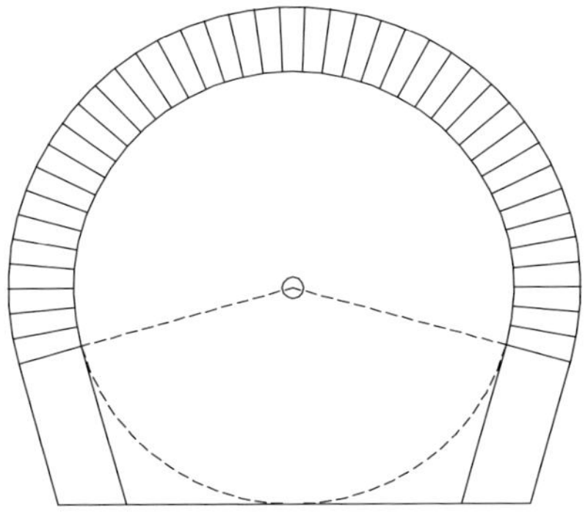

2

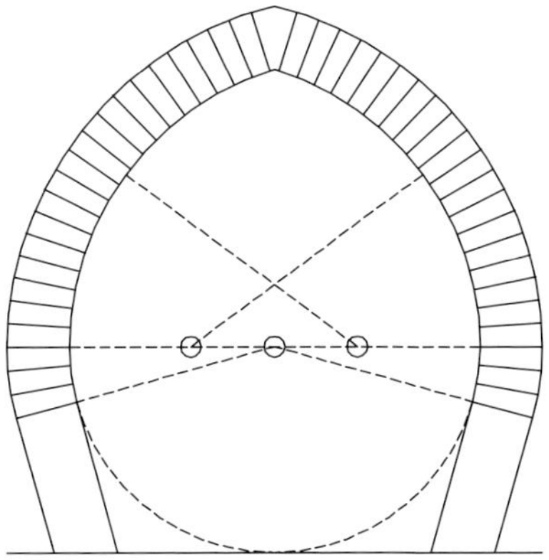

3

1a Semicircular arch, round arch, Roman arch

1b Semicircular arch - 2 piece

2 Round horseshoe arch, Arabic arch, Moorish arch

3 Pointed horseshoe arch

True arches

In the True arch the central or keystone took the load of the superstructure and shifted it sideways and gradually the load was transferred onto the post. The new technique used smaller pieces of stone and relied on limestone or brick and limestone mortar. The stones that formed the arch were not wedged inside the wall as in the case of the corbelled arch but were stand-alone wedges that sprung from the posts and were held together because of the keystone.

Thus, a True arch is composed of wedge-shaped blocks (typically of a durable stone), called voussoirs, with a keystone in the center holding them into place. Weight is transferred from one voussoir down to the next, from the top of the arch to ground level, creating a sturdy building.

1. Semicircular arch, round arch, Roman arch A semicircular arch that diverts pressure, through use of a keystone, down into the ground.

2. Round horseshoe arch, Arabic arch, Moorish arch These are a type of arch based on arcs of circles that has the centerline of the circle(s) higher than the springing line, making the body of the arch wider than the narrowest part of the opening.

Islamic architecture makes use of the semicircular arch to good effect, although a feature that defines Islamic buildings is the use of arches that use more than half a circle.

3. Pointed horseshoe arch Like the Gothic equivalent the pointed arch above is based around two arcs of circles. Once again the center line of the circles are above the springing line giving the distinctive Moorish look.

Above Doorway at Scarpa's Museo di Castelvecchio, Verona, Italy. Image: Rui Alves on Unsplash.

Right Tashkent, Uzbekistan. Image: AXP Photography on Unsplash.

4. Segmental arch, segmented arch, scheme arch The segmental arch was first developed by the Romans. In the twentieth century, this sort of arch was commonly used in the installation of doors and windows in private homes. The segmental arch—when the arc of a circle is smaller than 180 degrees—has a reasonable capacity to resist thrust and is among the strongest of arches.

In some cases, it may be necessary to incorporate rough axed arches, which are arches made of roughly cut hard bricks. For the sake of appearance, these bricks are cut and rough trimmed to fit the wedge shape of the voussoirs. Rough axed arches are not frequently used in modern construction techniques due to the extra cost, preparation time and level of skill required. It is also extremely difficult to cut axed arches from certain modern hard brick materials.

Because it forms part of a circle and is composed of half-brick rings, the segmental arch is very simple to build. A half-brick ring is a row of bricks showing heading faces only. The term 'rough' in this context means that the bricks used in the arch are not cut or trimmed to any shape or pattern but are simply left in their original condition. This means they present their rectangular ends to view. The flatter segmental shape has long been used for bridges. With the segmental arch the flatter it gets the more its thrust is delivered sideways to the abutments or embankments (sometimes referred to as 'impost') at each end of the bridge. With a bridge it is easy to build these abutments in a massive and solid manner to resist the sideways thrust.

5. Surbased arch A Gothic or pointed arch having a rise of less than half the span. As the center points of the arcs drop below the springing-line then the arch becomes flatter. Because it still has two separate arcs they still meet at a point, but as the arc centers approach the centerline of the opening the arch starts to look like a segmental arch.

6. Three-centered arch, basket handle arch, anse de panier The three-centered arch is based on two equal smaller arcs at the abutment-haunch area blending into a single larger radius arc at the crown. Also known as a Basket Handle or Anse De Panier.

7. Five-centered arch, elliptical arch An arch having the shape of half an ellipse. In its construction, the ellipse is often approximated by three adjoining circular arcs.

8. Rampant arch, raking arch An arch in which the impost on one side is higher than that on the other.

9. Parabolic arch An arch similar to a three-centered arch but whose intrados (the inner curved surface) is parabolic, with a vertical axis. In structures, their curve represents an efficient method of load, and so can be found in bridges and in architecture in a variety of forms.

10. Equilateral arch, three-pointed arch An arch described by two circular curves intersecting at the peak of the arch, each curve having a chord equal to the span. This form of the Gothic pointed arch in windows and arches was typically based upon an equilateral triangle in which the three sides have an equal length. This had the great advantage of simplicity.

11. Lancet arch, Acute arch The Lancet arch or Acute arch is a narrow pointed arch based on two segments of a circle, usually of the same radius. Thus, a Gothic or pointed arch is based on an equilateral triangle and has a rise greater than the span.

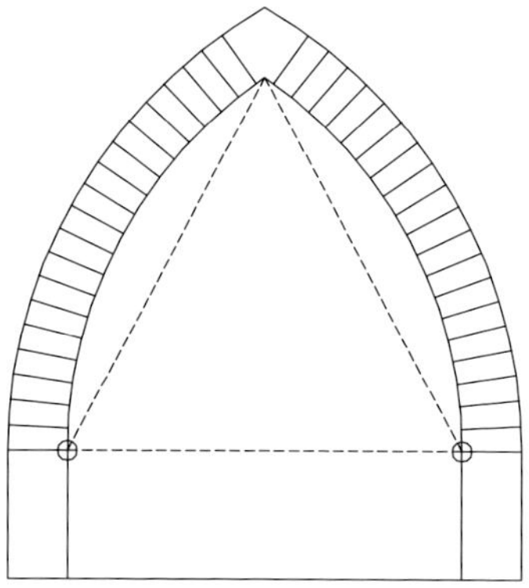

10

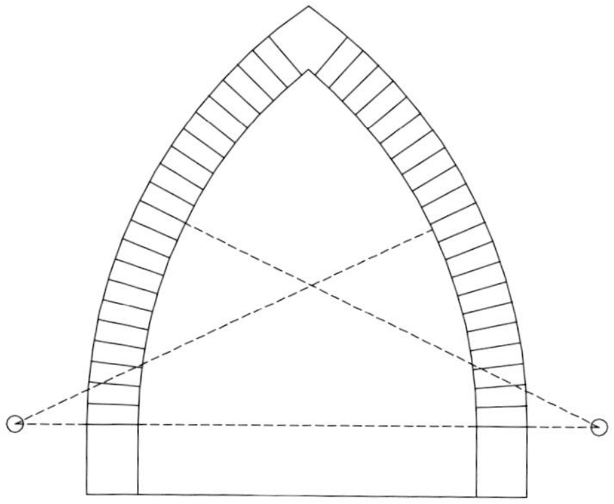

11a

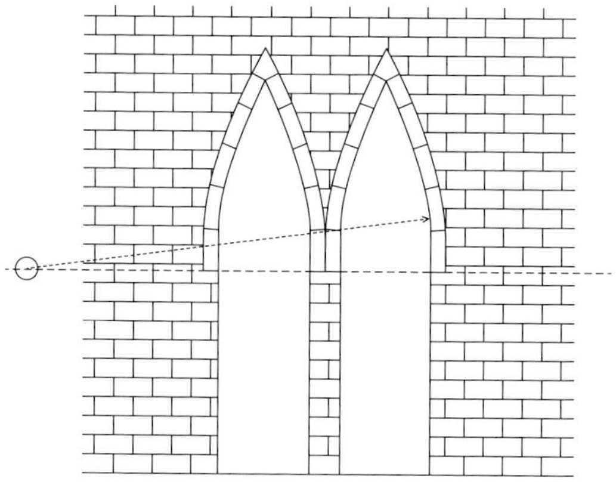

11b

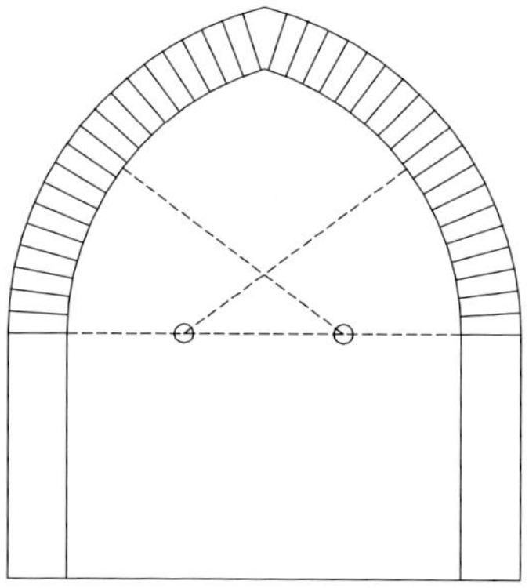

12

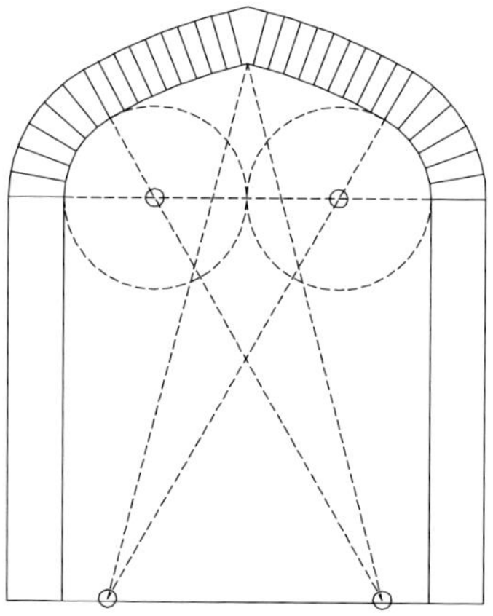

13

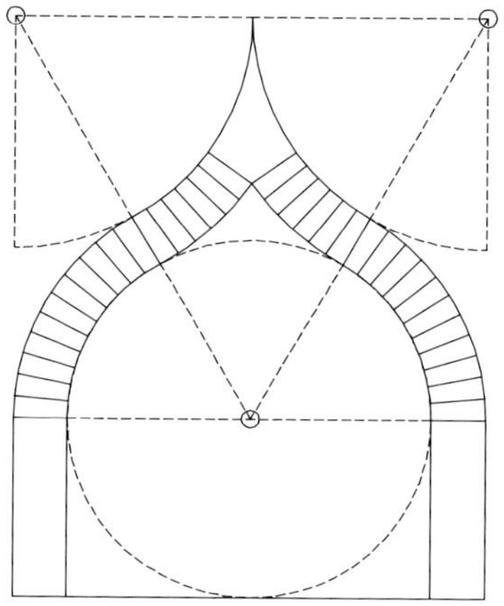

14

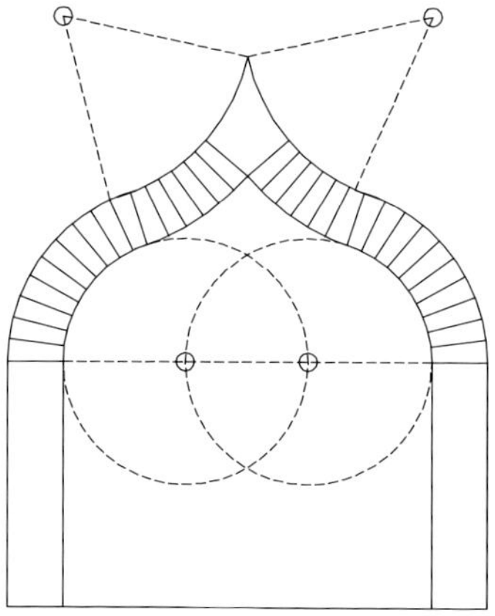

15

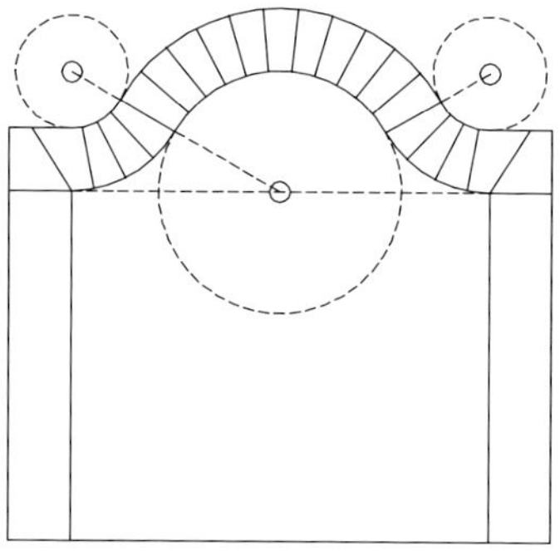

16

10 Equilateral arch, three-pointed arch

11a Lancet arch, acute arch

11b Lancet arch, acute arch

12 Drop arch, depressed arch

13 Four-centered arch, pointed segmental arch, Tudor arch

14 Ogee arch, inflected arch

15 Two-centered ogee arch, depressed ogee arch

16 Reverse ogee arch, bell arch

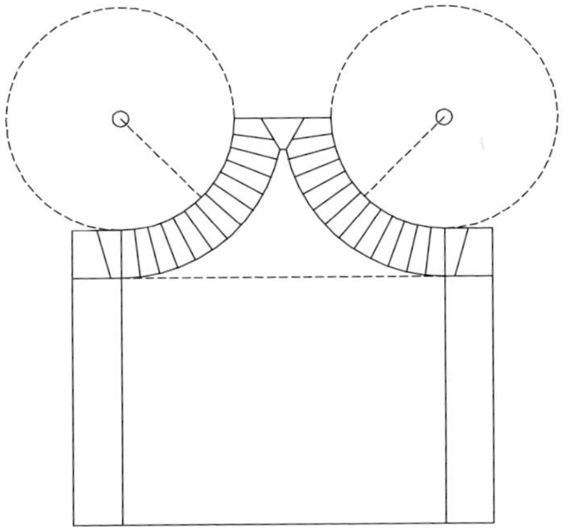
17a

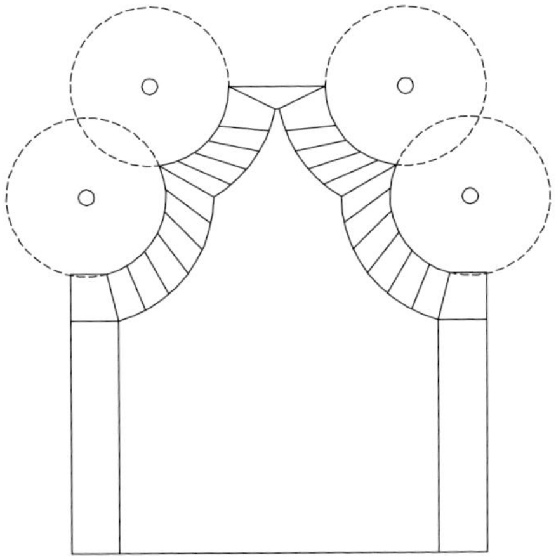
17b

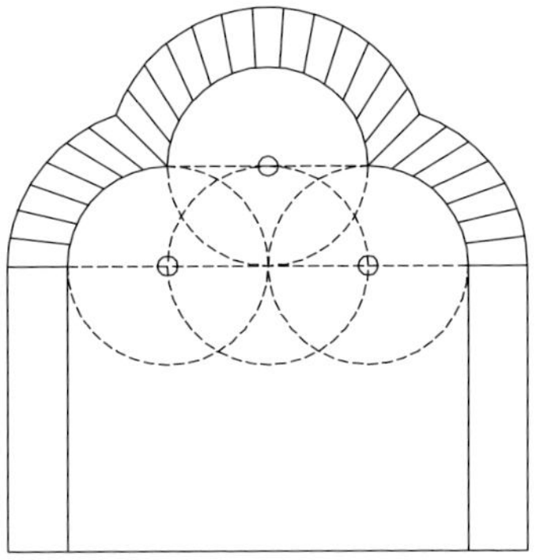
18

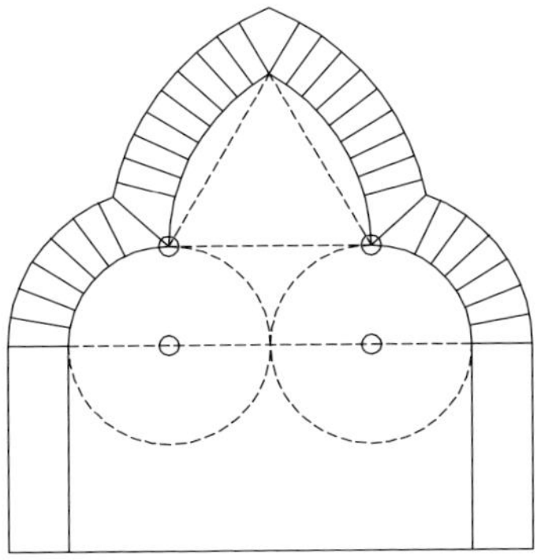
19

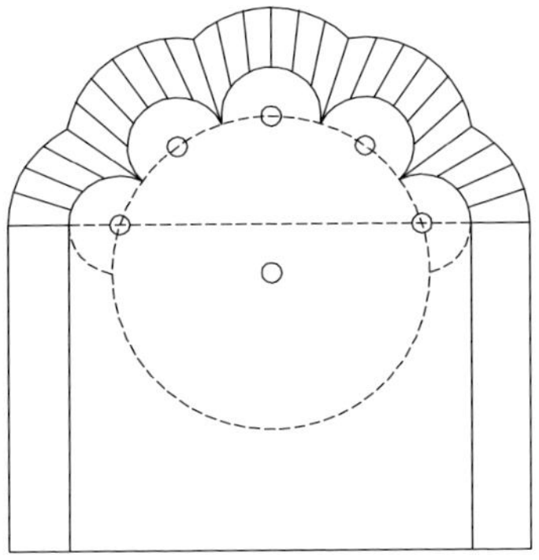
20

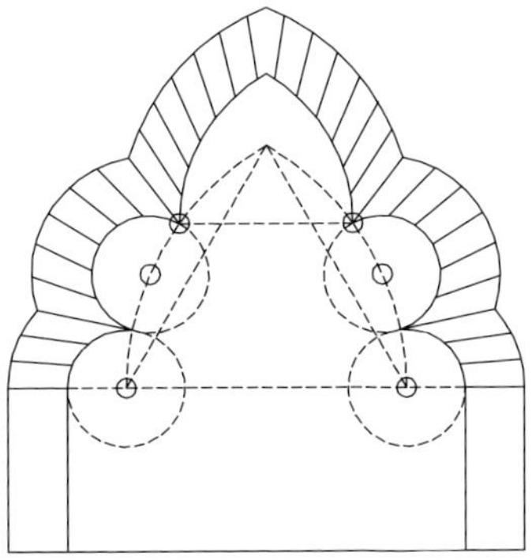
21

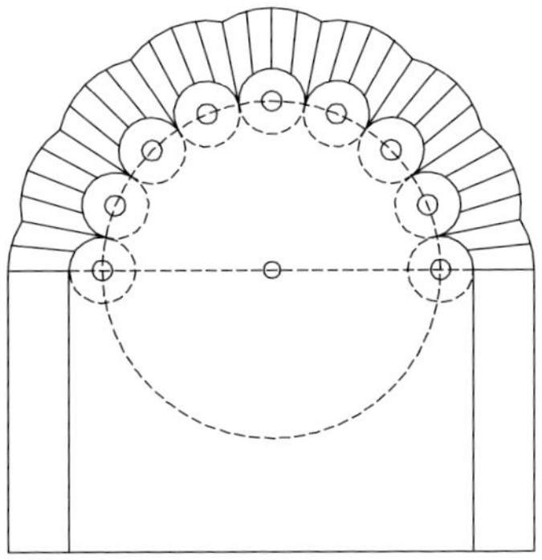
22

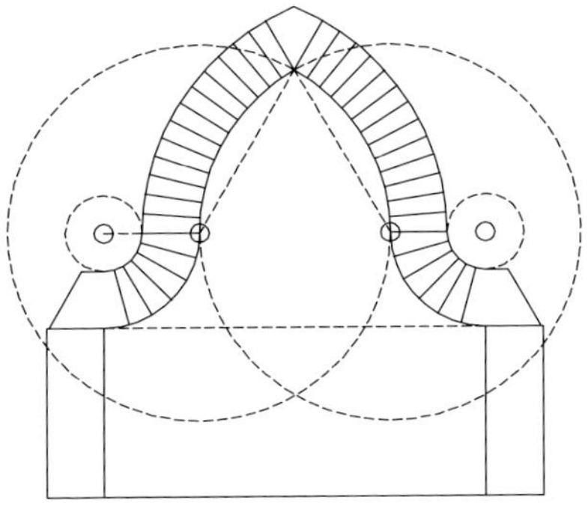
23

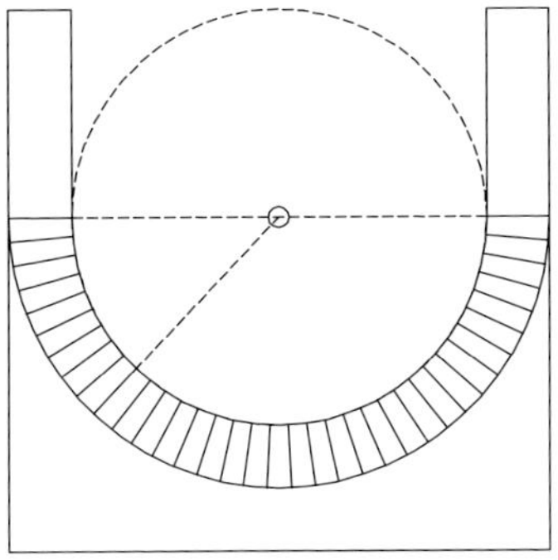
24

17a Tented arch, draped arch
17b Tented arch, draped arch
18 Trefoil arch
19 Pointed trefoil arch
20 Cinquefoil arch
21 Pointed cinquefoil arch
22 Multifoil arch
23 Keel arch
24 Inverted arch

Left top Arch detail, Morocco. Image: Annie Spratt on Unsplash.

Left bottom Cambridge Eco-friendly Mosque. Image: Rumman Amin on Unsplash.

Above Hallgrímskirkja, Lutheran church by Guðjón Samúelsson, Reykjavík, Iceland. Image: Hari Nandakumar on Unsplash.

Conclusion

This feature article has described the immense contribution made by arches to the world's greatest architectural monuments . Without the load-bearing capability of the arch, it is doubtful they could have ever been built.

For many thousands of years, public imagination has been in thrall to arches which have been the basic structural support for many buildings. Perhaps their appeal owes much – then and now – to the fact that for the builder a circle is simple enough to draw, and hence an easy way to set out an arch?

Many of the examples in this article explore the historical usage and importance of arches, whether in brick masonry or stone, which should come as no surprise. For hundreds of years the arch was a key component that supported vaulted cathedral ceilings.

Today's architects and designers look for innovative ways to transform the arch from the purely structural to the decorative and stylistic. They are alive to the possibilities the arch can offer, knowing that modern architecture has found other means of support using concrete, steel, wood and glass.

Instantly recognizable as a form, the arch has managed to seamlessly incorporate itself into countless modern projects, making it just as relevant today as it was in times past, regardless of the building material used.

Left Heydar Aliyev Cultural Centre, Baku, Azerbaijan.
Image: İltun Huseynli on Unsplash.

Brick shapes and brick types

Ron Ringer

Brick's continuing popularity with architects and builders lies with its ability to adapt to design and performative criteria. Brick can be shaped and used in a wide variety of traditional and contemporary details – from arches to sills, copings, columns and treads. This chapter is indicative of the many types of brick shapes available in North America, however, it should not be regarded as definitive.

Architects and designers often require shapes that don't fall within standardized design, and the availability of stock brick shape(s) required for a project will vary depending on the manufacturer. Glen-Gery is just one of a number of manufacturers in the United States that produce a wide range of brick shapes to match its own line of extruded, molded and handmade brick. Like most other manufacturers, it can produce customized shapes that conform to designer specifications.

Standard brick sizes

Standard brick sizes vary from country to country (a few examples below), with dimensions written as depth x height x length (D x H x L). Dimensions may be expressed in either imperial units (inches) or metric units (millimeters).

Country	Standard Brick Size in Inches (D x H x L)
USA	3 ⅝ x 2 ¼ x 7 ⅝
UK	4 x 2 ½ x 8 ½
Australia	3 x 4 ⅓ x 9
India	4 ¼ x 2 ½ x 9

Specified, actual and nominal dimensions

Three kinds of dimensions may be referenced:

- Specified dimensions – the anticipated dimensions of the brick itself, apart from the mortar. These are used in the project specifications, purchase orders, and when working with non-modular bricks.
- Actual dimension – the dimensions of a brick once it's manufactured, i.e. the actual end product.
- Nominal dimensions – typically expressed in round numbers (no fractions), are the sum of the specified dimensions and the expected thickness of the mortar.

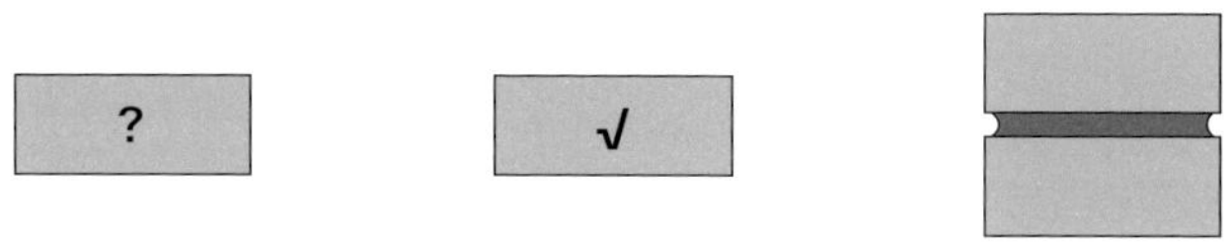

Specified Dimensions
anticipated manufactured dimensions

Actual Dimensions
actual measurements once manufactured

Nominal Dimensions
specified dimensions + anticipated thickness of mortar joint

Modular bricks

In new construction and renovation projects, components of a home or a business/industrial building are typically designed to fit together like pieces of a puzzle. That means the surrounding brickwork needs to be predictable in attributes such as shape, size, and load-bearing weight for the sake of efficiency and cost.

The nominal dimensions of modular bricks are sized either in round numbers or will add up to round numbers when bricks are grouped. Their standard and predictable sizes make it easy for them to be assembled together in construction, or in renovations, where they can be substituted for damaged or missing bricks.

Modular bricks have specified, actual, and nominal dimensions, whereas non-modular bricks only have the first two, and lack nominal dimensions.

The most common mortar joint size is ⅜ of an inch, or .38 inch, as specified by the International Building Codes TMS 602, on the *Specifications for Masonry Structures (TMS 602-16).*[1] Mortar joints that measure ½ of an inch are also common.

Note on coring: For all manufacturers, coring of standardized brick shapes is a plant/manufacturer option. Any coring requirements to these shapes place them into a custom shape category, requiring drawings.

Modular bricks

Sized so that the nominal dimensions fit into multiples of 4 inches

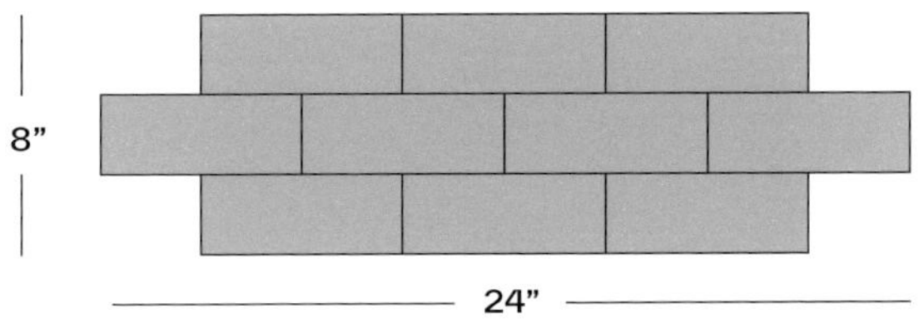

Nominal dimensions of a modular brick (D x H x L): 4” x 2 ⅔” x 8”

Mortar joint thickness ⅜”

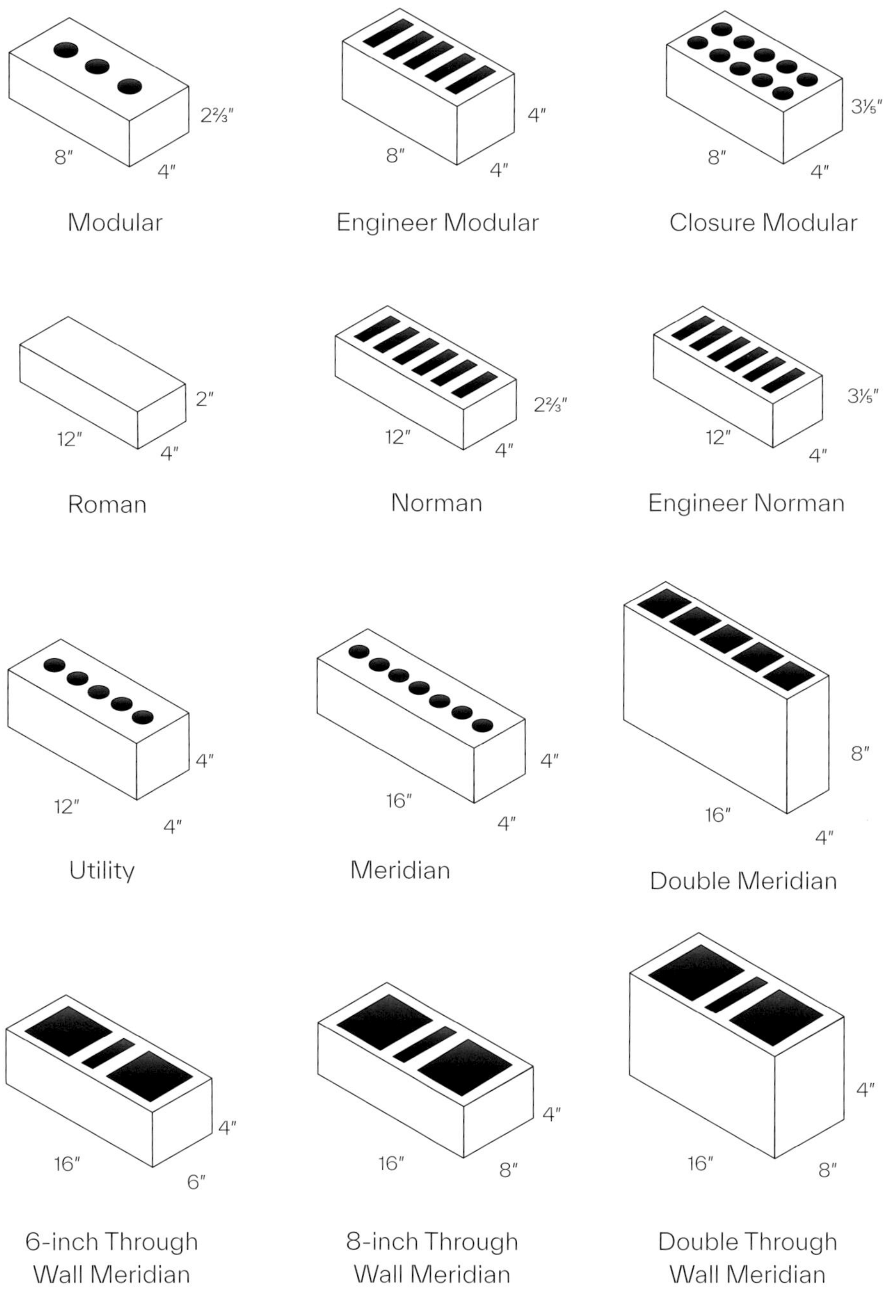

Brick Type	Description	Mortar Joint Thickness (inches)	Specified Dimensions (D x H x L) (inches)	Nominal Dimensions (D x H x L) (inches)
Modular	A nominal measurement in modular brickwork incorporates not only the specified (manufacturer anticipated) dimensions, but also the recommended thickness of the mortar that will connect it to other bricks. A structural modular brick will have two or three holes pierced completely through the center and no raised edges for easier planning on the part of builders and masons.	⅜	3 ⅝ x 2 ¼ x 7 ⅝	4 x 2 ⅔ x 8
(i) Closure Modular		⅜	3 ⅝ x 3 ⅝ x 7 ⅝	4 x 4 x 8
(ii) Closure Modular	These types of brick are pierced with more holes than traditional modular brick. With more hollows/holes for the mortar to 'grab onto' built into the brick, these are used to finish corners and edges, hence their name.	½	3 ½ x 3 ½ x 7 ½	4 x 4 x 8
Engineer Modular	Engineer modular bricks use rectangular slots rather than round holes for the brick piercing and are more commonly used for residential projects like homebuilding. These holes through the brick are created to assist in manufacturing, allowing more even cooling of the brick material during creation and lighter weight for shipping and handling during construction.	⅜	3 ⅝ x 2 13/16 x 7 ⅝	4 x 3 ⅕ x 8
Engineer Modular		½	3 ½ x 2 ¾ x 7 ½	4 x 3 ⅕ x 8
Jumbo	The jumbo brick is simply larger in size. Hard and compressed, it is used for adding structural strength to buildings, arches, foundation, pavements and bridges. Alternative names include antique jumbo brick, old jumbo brick, used jumbo brick.	⅜	3 ⅝ x 2 ¾ x 8	4 x 3 x 8
Jumbo		½	3 ½ x 2 ½ x 8	4 x 3 x 8
Roman	Named for their area of origin, Roman bricks are flat and slender. They are used to deliver a distinct and attractive finish for building fronts, columns, and other areas where aesthetics are valued on equal footing with structural integrity. Their smooth appearance, typically unmarked by rough finishes or piercing, is memorable and pleasing to the eye.	⅜	3 ⅝ x 1 ⅝ x 11 ⅝	4 x 2 x 12
Roman		½	3 ½" x 1 ½" x 11 ½	4 x 2 x 12
Norman	Like Roman bricks, the Norman style is named after the first group of historical people to have used it. Very similar in size to the Roman brick. The difference between the two, other than height, is that Norman bricks have rectangular piercings for mortar flow-through, whereas Roman bricks do not.	⅜	3 ⅝ x 2 ¼ x 11 ⅝	4 x 2 ⅔ x 12
Norman		½	3 ½ x 2 ¼ x 11 ½	4 x 2 ⅔ x 12

Brick Type	Description	Mortar Joint Thickness (inches)	Specified Dimensions (D x H x L) (inches)	Nominal Dimensions (D x H x L) (inches)
Engineer Norman	Building quite literally on the height of the Norman-style brick, the Engineer Norman variety adds a bit more material to work with. Like its shorter namesake, this brick also features pierced slots for the mortar to grip, ensuring a better overall hold as the courses are built atop one another.	⅜	3 ⅝ x 2 13⁄16 x 11 ⅝	4 x 3 ⅕ x 12
Engineer Norman		½	3 ½ x 2 ¾ x 11 ½	4 x 3 ⅕ x 12
Utility	The utility modular brick is similar to the base modular brick in several design concepts – plain-sided, pierced with holes that are round – with the key differences found in overall height and depth. These bricks are particularly easy to match with other building components like windows and door frames.	⅜	3 ⅝ x 3 ⅝ x 11 ⅝	4 x 4 x 12
Utility		½	3 ⅝ x 3 ⅝ x 11 ⅝	4 x 4 x 12
Meridian	For building purposes that require more individual brick length than the utility brick can provide, the meridian adds a third more length. The extra four inches of length over the utility style allows for more flexibility and stability in offset patterning, while the pierced holes ensure good mortar grip.	⅜	3 ⅝ x 3 ⅝ x 15 ⅝	4 x 4 x 16
Meridian		½	3 ½ x 3 ½ x 15 ½	4 x 4 x 16
Double Meridian	In building scenarios where more of both height and length are needed in the individual brick sizing, the double meridian brick offers a solution. The 'double' refers to the height over the standard meridian's 4 inches. The cores on this brick are also considerably larger and squared off rather than rounded, making the brick a 'hollow' construction brick.	⅜	3 ⅝ x 7 ⅝ x 15 ⅝	4 x 8 x 16
Double Meridian		½	3 ½ x 7 ½ x 15 ½	4 x 8 x 16
6-Inch Through Wall Meridian	This hollow brick style is half the height of a double meridian, yet builds on the style with an extra 2 inches of depth. The large squared-off cores allow easier handling for masons, delivering the outer durability of brick without the weight demands of more solid brick with smaller, circular cores.	½	8 x 6 x 16	6 x 4 x 16
8-inch Through Wall Meridian	Similar to 'solid' brick styles, hollow meridian bricks come in several sizes that can be overlapped and combined into virtually any modular brick building project.	½	8 x 4 x 16	8 x 4 x 16
Double Through Wall Meridian	A hefty, yet hollow, brick style that gives builders a substantial modular option for large jobs with substantial demands on coverage in courses.	½	8 x 8 x 16	8 x 8 x 16

Non modular bricks

Non-modular brick sizes are not standard and do not fit easily into a conventional pattern or structure, e.g. around window openings or doors. They are best suited to unconventional builds that may call for odd sizes. These bricks have specified and actual sizes, but not nominal dimensions. They are usually available in different sizes.

Larger bricks are cheaper which means that using non-modular bricks saves money. For example, a wall made of non-modular king brick can be 25 percent less expensive than the cost of a wall constructed from modular brick. Note that larger bricks are also often used as facing bricks and may not have the same structural capabilities. Also, heavier bricks can be more difficult for the bricklayer to work with, because it's common for them to hold bricks with one hand, facilitating the building process.

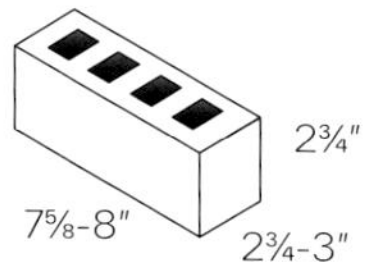

Queen

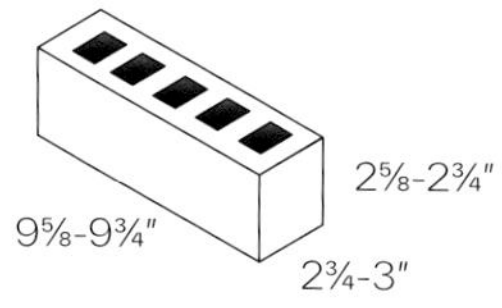

King

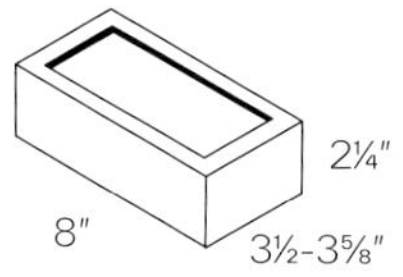

Standard

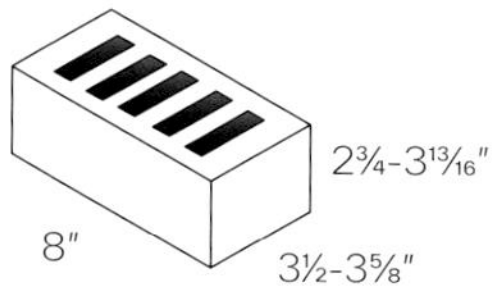

Engineer Standard

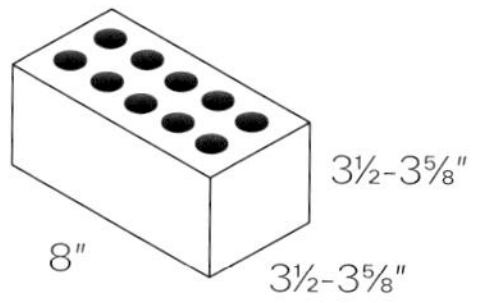

Closure Standard

Brick Type	Description	Mortar Joint Thickness (inches)	Specified Dimensions (D x H x L) (inches)
King	A non-modular king brick is larger than its queen counterpart, with an additional squared core built into a more substantial length.	⅜	2 ¾ x 2 ⅝ x 9 ⅝
King		½	3 x 2 ¾ x 9 ¾
Queen	A non-modular queen brick has large square cores for mortar flow through. Because this commonly-used brick format has evolved and changed with the history of masonry, the sizes of a queen brick can vary slightly between manufacturers and will typically be called out in sales information if this is the case.	⅜	2 ¾ x 2 ¾ x 7 ⅝
Queen		½	3 x 2 ¾ x 8
Standard	A standard non-modular brick incorporates a design element known as a frog – a divot in the center of the brick that gives it a shallow tub-like appearance. It serves the same purpose as cores do, giving the mortar a place in which to flow and 'grip' as it dries.	⅜	3 x 2 ¾ x 8
Standard		½	3 ½ x 2 ¼ x 8
Engineer Standard	Very similar in appearance and design to the modular engineer brick, this style features the same five rectangular core placements and varies only in measurements.	⅜	3 ⅝ x 2 ¹³⁄₁₆ x 8
Engineer Standard		½	3 ½ x 2 ¾ x 8
Closure Standard	The closure standard brick is a close cousin of its modular counterpart. Built with the same circular cores and nearly the same dimensions, it's also used to finish corners and edging with its clever flow-through design that offers a firm mortar grip.	⅜	3 ⅝ x 3 ⅝ x 8
Closure Standard		½	3 ½ x 3 ½ x 8

Square brick options

Square bricks, also known as oversize bricks, are designed to be used in large-scale institutional building projects such as schools, office buildings, and hospitals. The additional height of the bricks gives masons the ability to create long, high walls without compromising overall stability. Long, deep cores that pierce through the entirety of the brick ensure that construction is built to last. Overall, the six styles of 8-Square bricks are designed to work together, with matching 7⅝ inch heights to build an even course that remains steady as wall height increases.

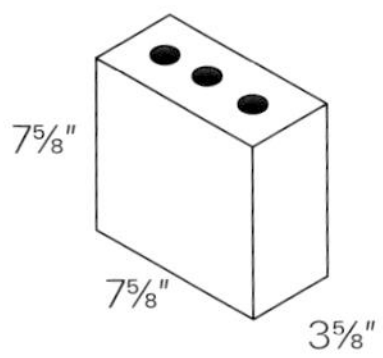

8SQ

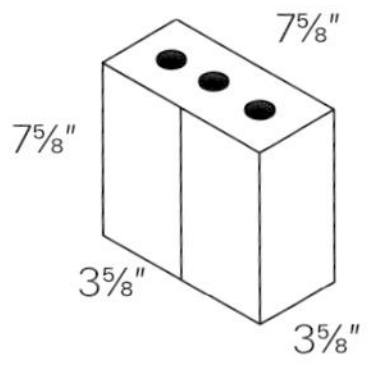

8SQ1

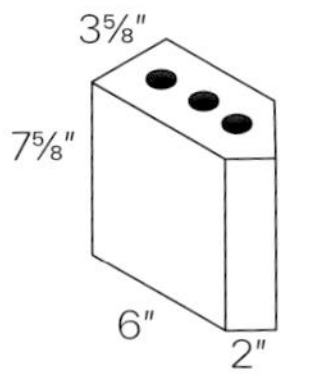

SQ External

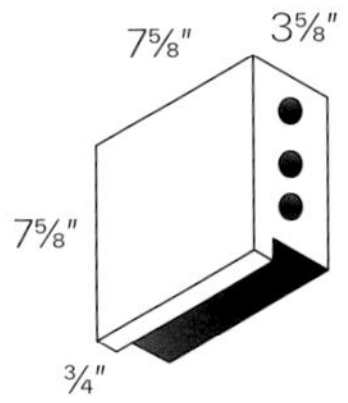

8SQ Stretcher - Header

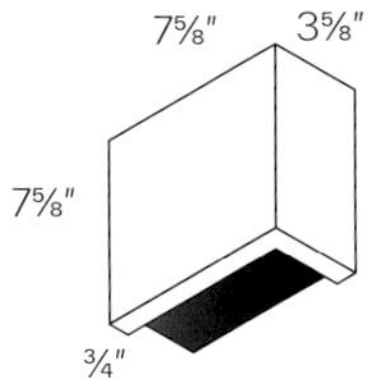

8SQ Corner Stretcher - Header

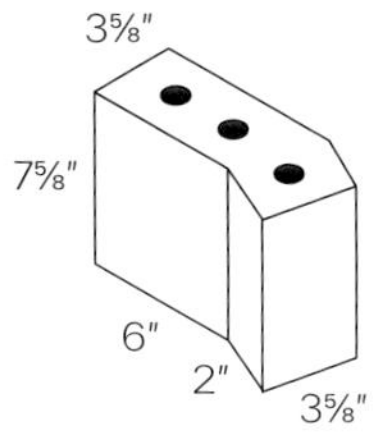

8SQ Outer - Inner

Brick Type	Description
8SQ 8SQ 1	The 8SQ and the 8SQ 1 are essentially the same brick in terms of overall specified dimensions – 3⅝ x 7⅝ x 7⅝ – the 8SQ1 also features scoring at the mid-point that visually bisects, via the front face, the area in the brick containing a central core hole.
8SQ External	The 8SQ External shortens the brick length by a little less than 2 inches, measuring 6 inches and adds a roof-like angle onto one edge, with each slant edge measuring 2 inches in length.
8SQ Stretcher-Header 8SQ Corner Stretcher-Header	The 8SQ Stretcher-Header and the 8SQ Corner Stretcher-Header are also virtually identical: both measure 3⅝ x 7⅝ x ⅝ with a ¾ inches in depth intended to overlap brick coursework. The key difference is that the 8SQ Corner Stretcher-Header omits the circular cores found through the 8SQ Stretcher-Header.
8SQ Outer-Inner	The 8SQ Outer-Inner finishes the group with a unique 'bent edge' design that can be used to build angular walls and borders. This brick measures 3⅝ x 7⅝ x 8 (6 inches length for the body, 2 inches depth for the angular offset).

Water Table bricks

So-called water tables were once a functional element in building design intended to reduce water penetration. Today, they are more commonly used to enhance visual appeal by adding depth and detail. These details can range from the use of a single style for simplicity to a combination of styles in order to achieve highly ornate brickwork. In addition to their traditional use, these shapes can be used to add dimension to flat walls, rotated to form door and window jambs, or inverted to form cornice brickwork. Water tables can be simple or elaborate to match the size and design of the building. Continuity of design can also be transferred into the landscape by using the water tables as caps for walls or edging accents for walkways.

On the flat Water table bricks (on the flat) are similar to coping bricks and are designed to direct water away, not only from the brickwork structure itself, but also from its base. Pooling water can cause long-term damage and weakening to both mortar and brickwork, compromising the stability and safety of the structure. So-called on the flat style water table bricks perform this duty but are mostly used as incorporation to add decorative elements to brickwork. When used to create columns and rows within a wall of traditional brickwork, they add both decorative textures and shadowing.

There are two types of on the flat water table systems based on the brick position in the wall. Some manufacturers can provide either/both types with companion exterior and interior corners. Water tables can also be turned upside down to achieve a cornice detail.

Types include: Water table in the header position (H); water table in the stretcher position (S).

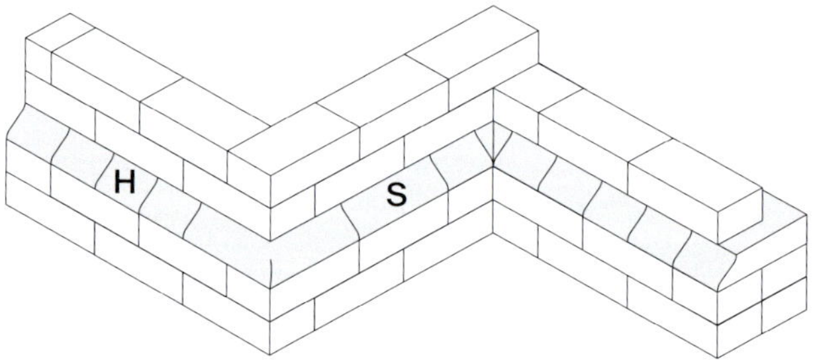

Water table bricks in the rowlock position Rowlock water table bricks are created with more function than form in mind. The notches, divots, and angular lip designs available in this group of bricks do still have plenty of style capabilities, but they also work hard to redirect water off of walls, sills, roofs, and more. Depending on the slope and size of their vertical stack within the coursing, this water-redirection may be a gentle effect or a dramatic one with a substantial distance away from the footing.

Water table in the rowlock brick position (R). Water table in rowlock position can be rotated to the header position (H). Water tables can also be turned upside down to achieve a cornice detail.

Following examples The majority of examples show water table brick systems manufactured by Glen-Gery.

Water tables on the flat

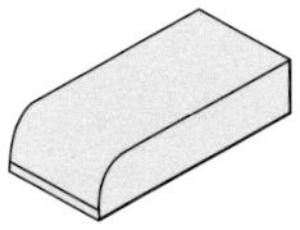
WT1 -
Bullnose header

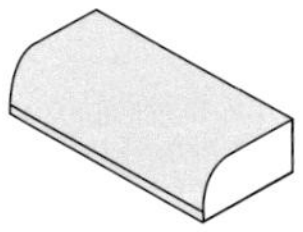
WT1 -
Bullnose stretcher

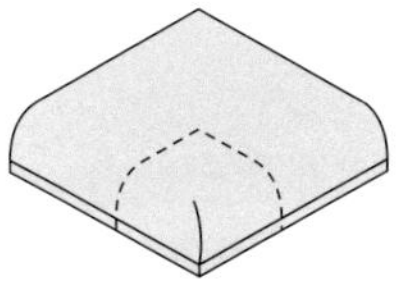
WT1 - Bullnose
header/stretcher external

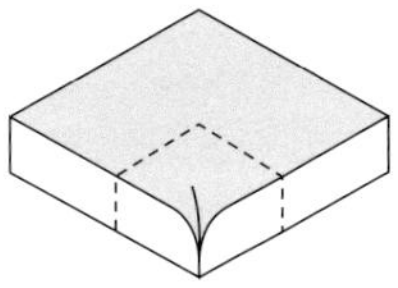
WT1 - Bullnose
header/stretcher internal

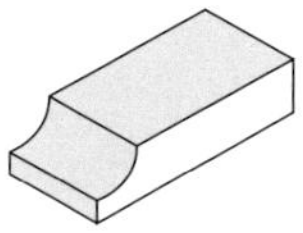
WT2 -
Cove header

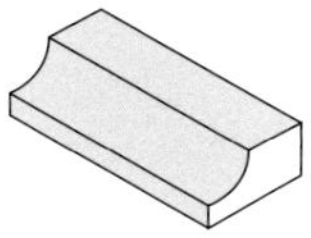
WT2 -
Cove stretcher

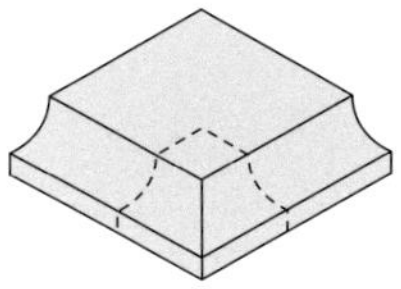
WT2 - Cove header/
stretcher external

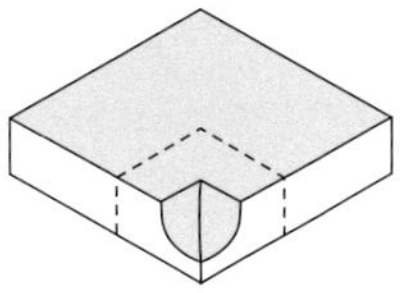
WT2 - Cove
Header/Stretcher internal

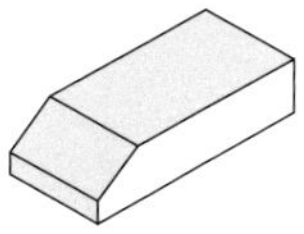
WT3 -
Beveled header

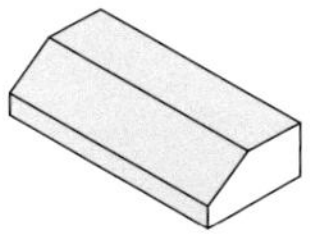
WT3 -
Beveled stretcher

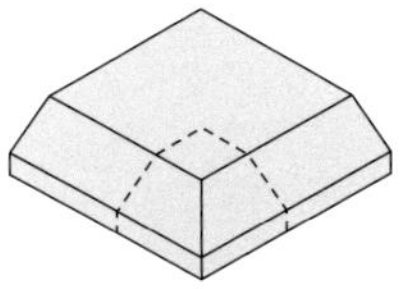
WT3 - Beveled header/
stretcher external

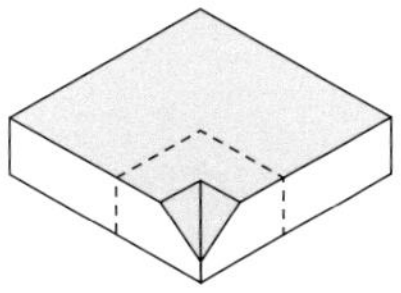
WT3 - Beveled
header/stretcher internal

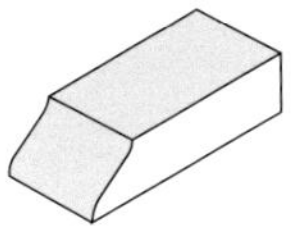
WT4 - Ogee
header

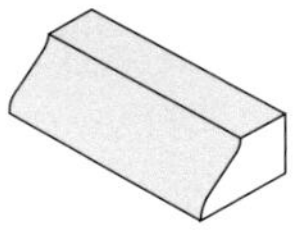
WT4 - Ogee
stretcher

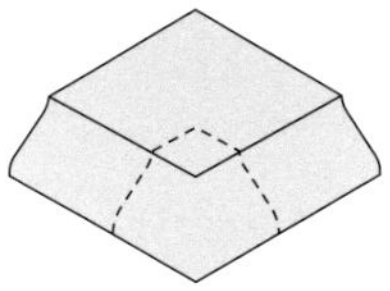
WT4 - Ogee header/
stretcher external

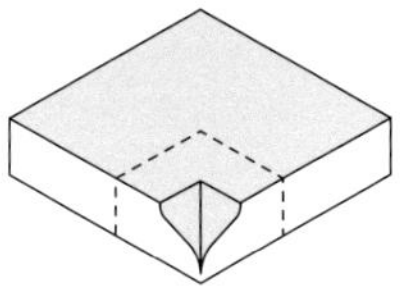
WT4 - Ogee header/
stretcher internal

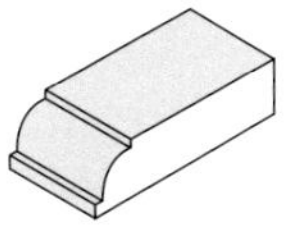
WT5 - Header

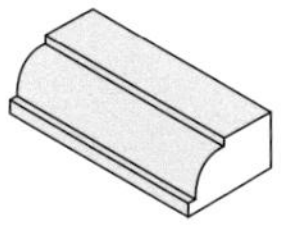
WT5 - Stretcher

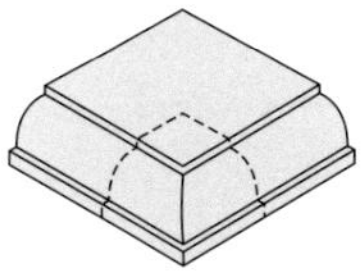
WT5 - Header/
stretcher external

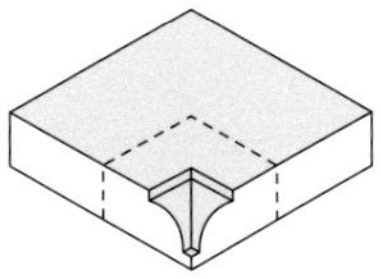
WT5 - Header/
stretcher internal

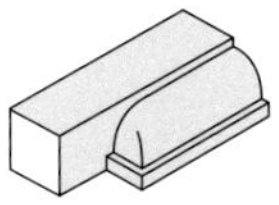
WT5 - Standard left
(shown) or right

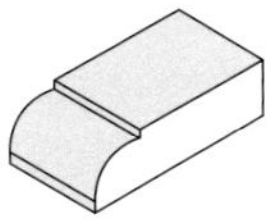
WT6 - Header

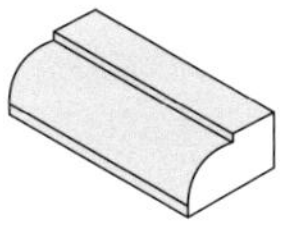
WT6 - Stretcher

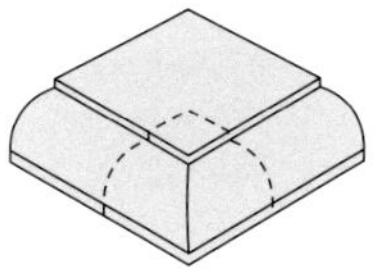
WT6 - Header/
stretcher external

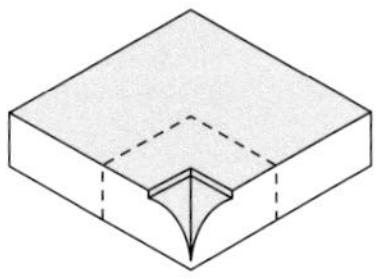
WT6 - Header/
stretcher internal

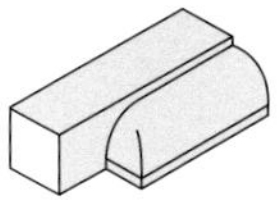
WT6 - Standard left
(shown) or right

Water tables rowlock (WTR)

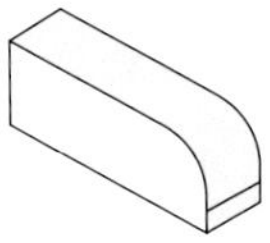
WTR1 - Bullnose header

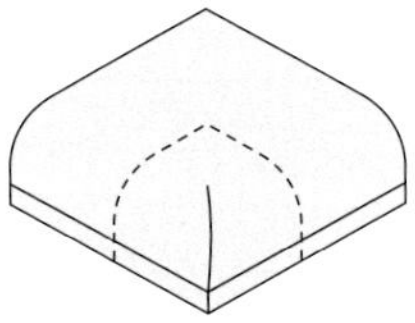
WTR1 - Bullnose corner external

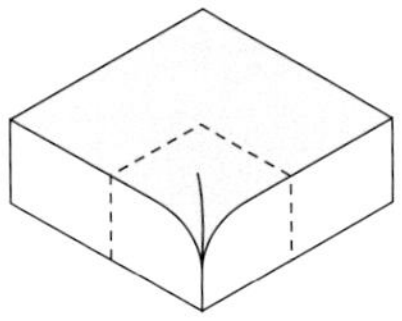
WTR1 - Bullnose corner internal

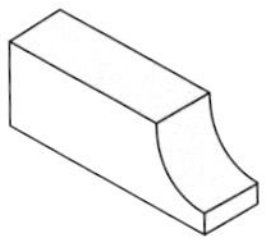
WTR2 - Cove header

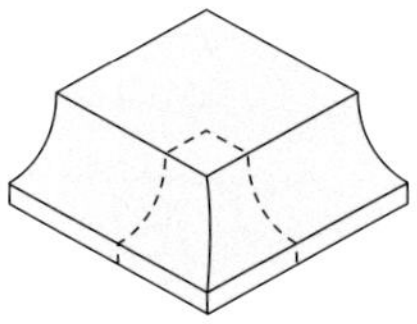
WTR2 - Cove corner external

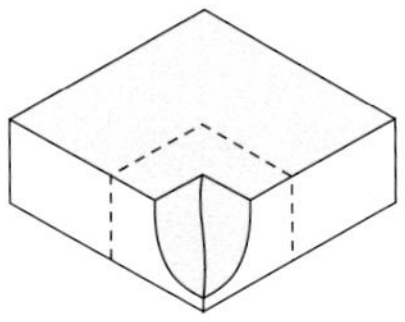
WTR2 - Cove corner internal

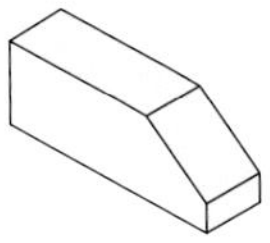
WTR3 - Beveled header

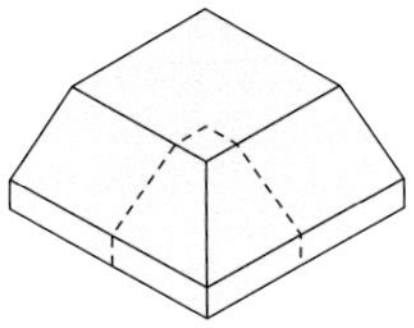
WTR3 - Beveled corner external

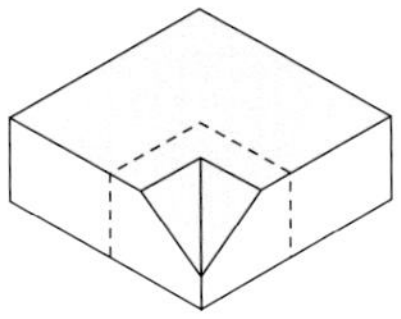
WTR3 - Beveled corner internal

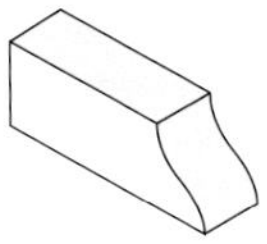
WTR4 - Ogee header

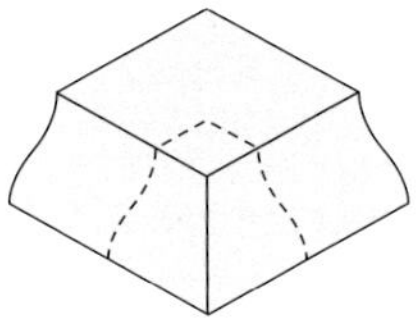
WTR4 - Ogee corner external

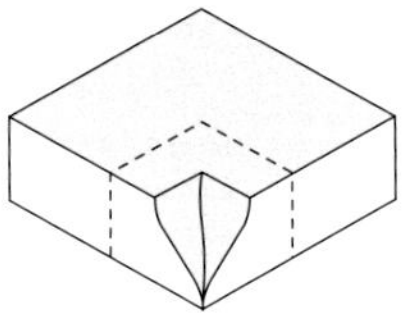
WTR4 - Ogee corner internal

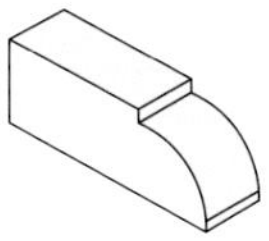
WTR5 - Header

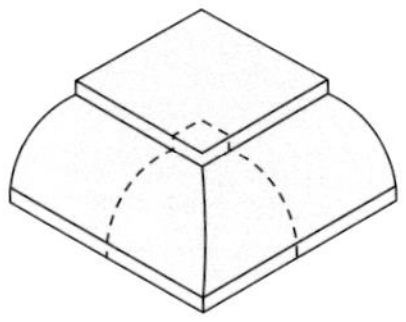
WTR5 - Corner external

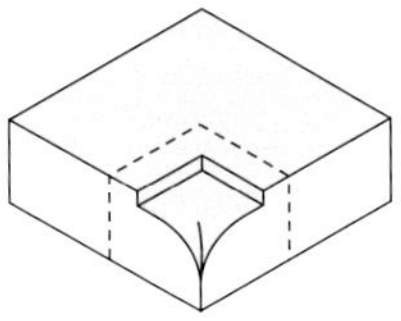
WTR5 - Corner internal

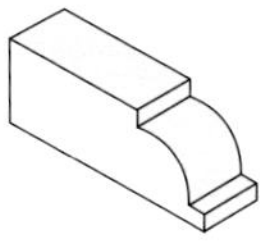
WTR6 - Header

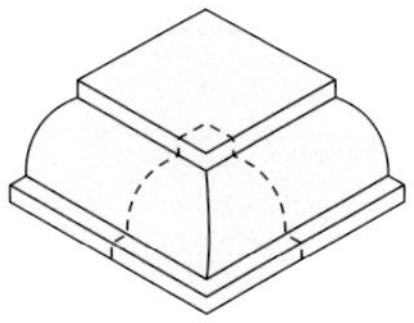
WTR6 - Corner external

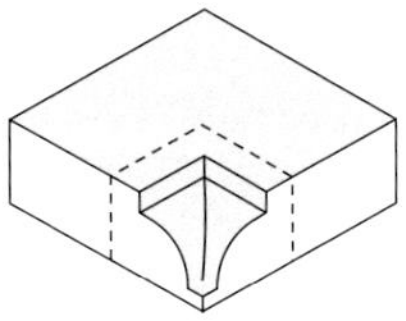
WTR6 - Corner internal

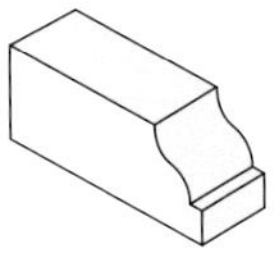
WTR7 - Header

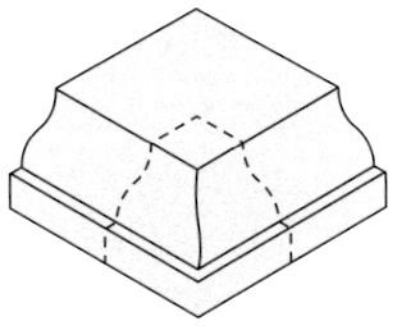
WTR7 - Corner external

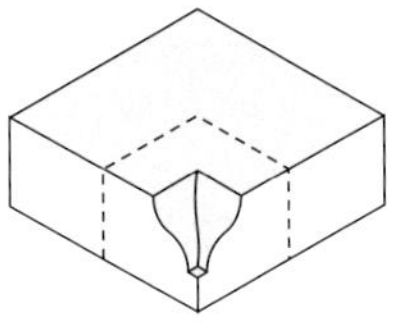
WTR7 - Corner internal

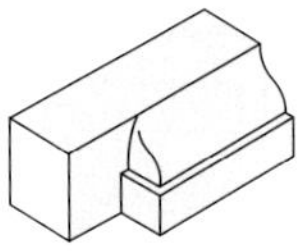
WTR7 - Internal

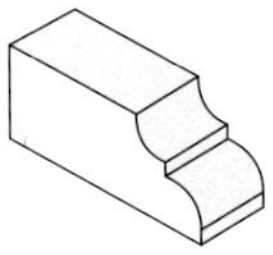
WTR8 - Header

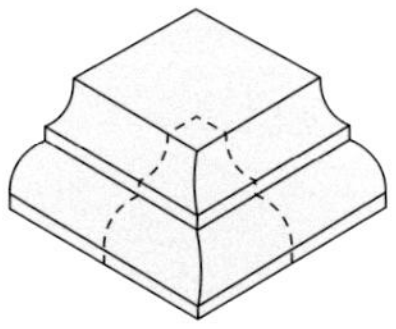
WTR8 - Corner external

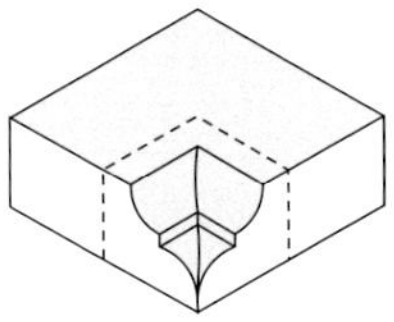
WTR8 - Corner internal

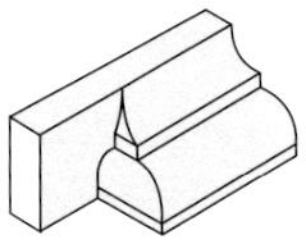
WTR8 - Internal

Coping or wall caps

Copings form a cap on top of a wall that protects the brick masonry below from water penetration The right type of coping keeps water from pooling on the brickwork itself and also directs it away from the base of the structure. Additionally, coping bricks offer a great deal of aesthetic quality when used decoratively.

Coping bricks come in varying shapes and sizes. These brick styles are not only crucial for protecting the mortared brick coursing underneath them, but they're also vital for moisture management as well. The degree of curvature (or lack thereof), height, and size of a coping will determine where water flows down structures during storms.

Bullnose wall cap bricks Bullnose bricks are often used to top off (cap) walls, giving them a soft, elegantly finished appearance. Bullnose bricks are also used for a rounded quoin (round corner) and as the rounded edge for staircase steps, building corners, verandas, etc. A brick which is manufactured with a double bullnose on end is known as a cownose brick and is also used to cap walls. Ogee caps and bell caps, for example, soften corners and angles. Half-moon caps give walls a rounded top, and ridge caps have straight sides that slope away from a sharp peak.

There are many different styles of bullnose-type bricks designed to for specific purposes. Types include: bullnose rowlock, bullnose header, bullnose stretcher, and bullnose water table bricks. It is important to remember that naming conventions vary with the manufacturer, and may change over time.

Coping or wall caps

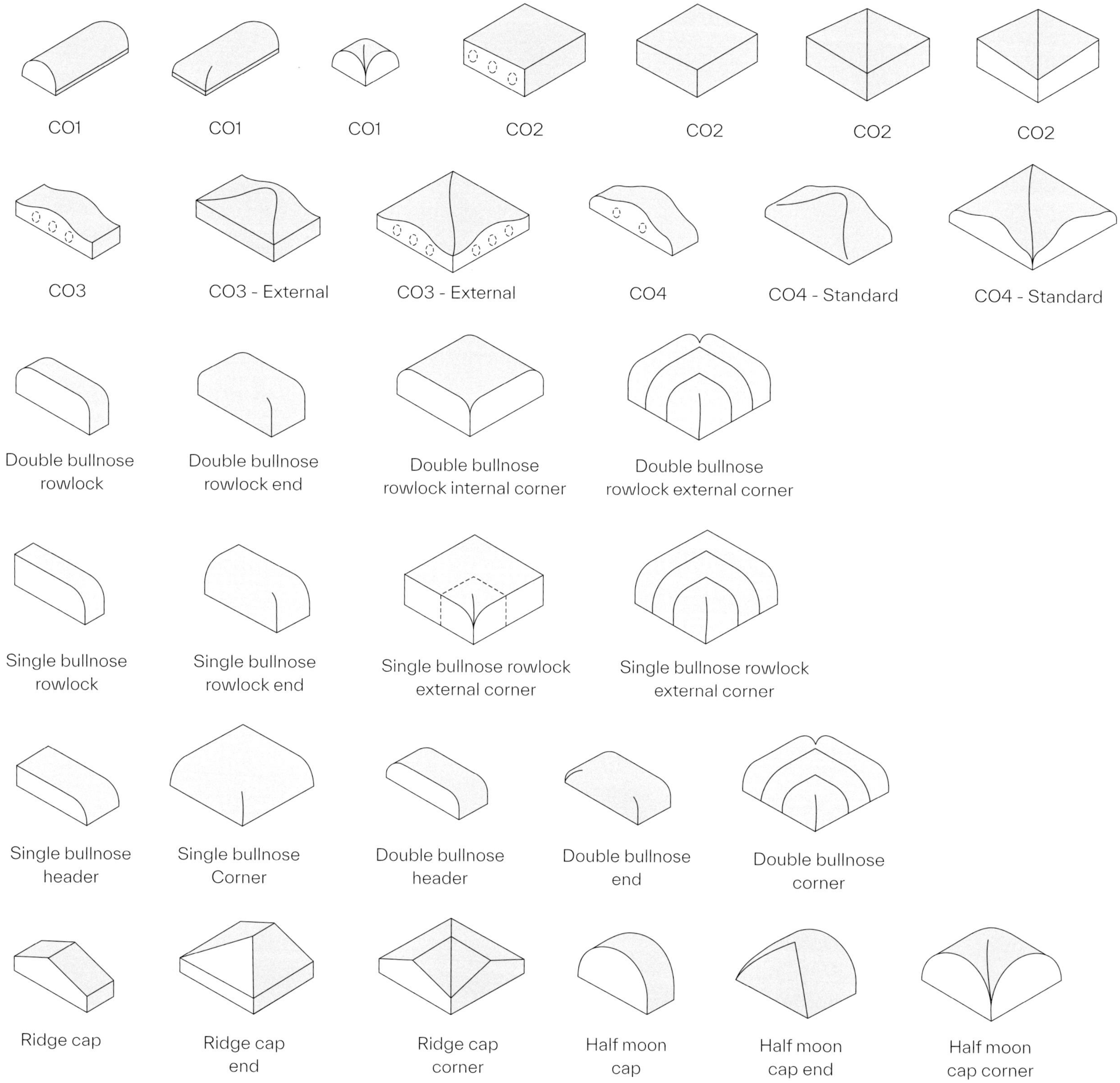

Sill and tread bricks

Sill bricks The term sill derives from the old English word syll, meaning 'style' and the Germanic word schwell, meaning 'threshold.' Sill bricks come in different styles, depending on the manufacturer. Sill bricks are invaluable for providing iconic looks on windows and doorways across the country. Their smooth, rounded edges and crisp angular options provide builders with a variety of options for considerations like water run-off and overall aesthetic appearance. In particular, sills are used to prevent water penetration below window and door openings. Their sloped design drains water away from these critical areas.

Tread bricks have relatively flat, smooth, and rounded-edge styles and are used in conjunction with conventional rectangular bricks to build outdoor stairs for homes and institutions alike. The comparatively short height of tread bricks ensures more even settling on mortar and other bricks, while the rounded edge offers the comfort of both the aesthetic and physical variety. Rounded-edge steps are more comfortable for sitting. Tread bricks may also be used to create pathways or pathway edges on properly-leveled, reinforced ground.

Ogee bricks give treads and sills a soft, smooth finish and a look of refined elegance. They speak of quality, gentility and the pride of craftsmanship. The smooth, rounded edges of ogee treads can turn outdoor steps into a gracious, inviting entrance. Ogee sills lend the same aesthetic enhancement to windows, giving them a firm foundation on which to rest. The soft edges and horizontal lines of ogee sills are accent features that add depth and definition.

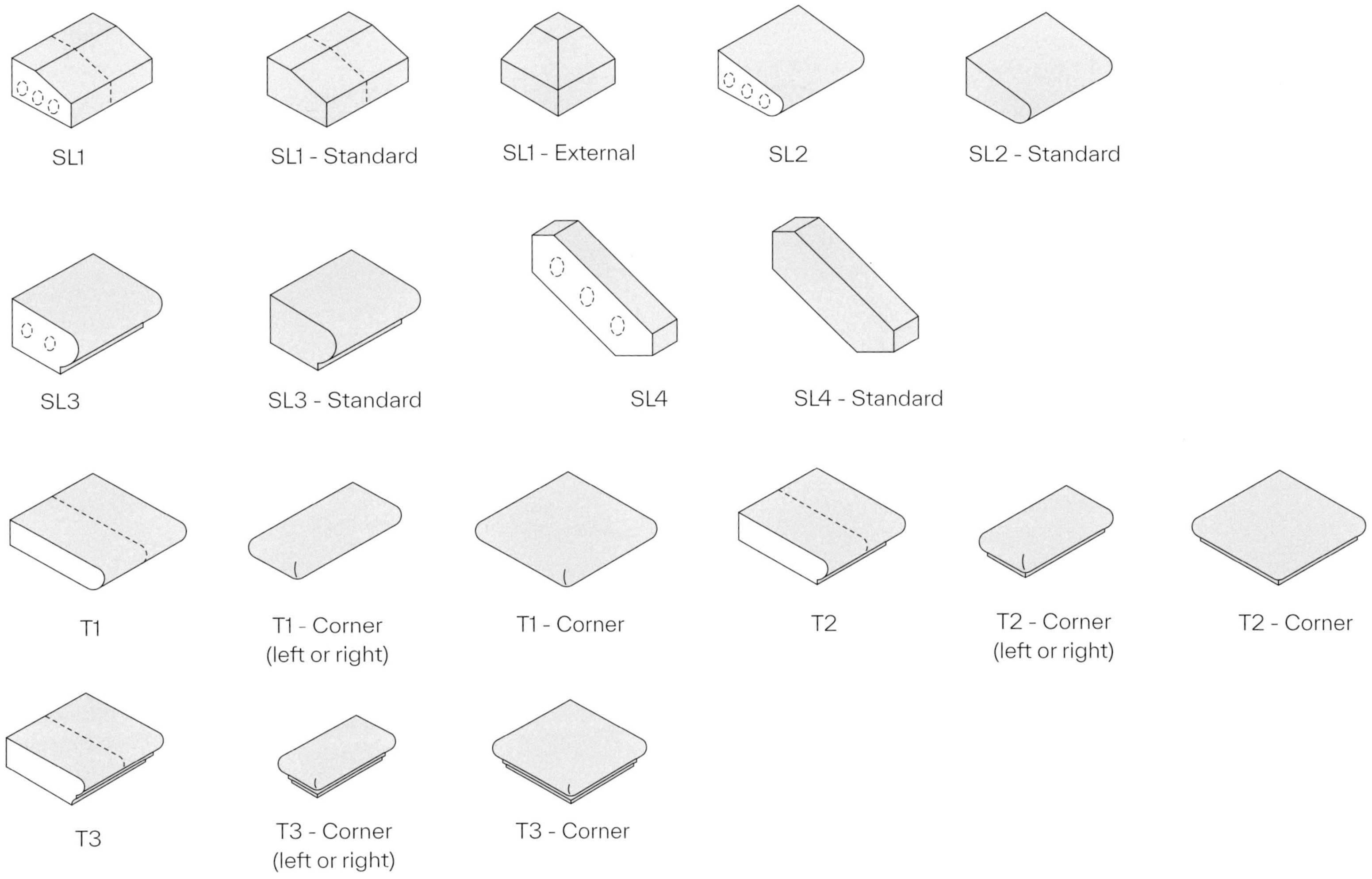

Corner bricks

While the rectangular or squared dimensions of most standard bricks can create straight border lines, variations in topography, structural stability, and even style require a need for variation. Corner bricks offer masons the flexibility to create anything from a one-off exception built into a wall to avoid obstruction to a planned octagonal (or other geometrically-inspired) shape in their brickwork. These work in tandem with standard bricks to remove the restrictions of straight-angle-only buildings while laying courses.

The octagon and hingebrick are designed to provide alternative corner treatments that do not require field cutting. The octagon brick shapes form 135° interior or exterior angles while maintaining coursing and visual continuity. The unique design of the hingebrick offers even greater design flexibility. It allows an unlimited range of angles at corners and can be adapted for use in other brickwork details.

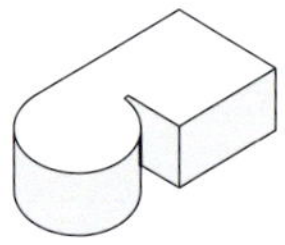

Hingebrick - corner brick

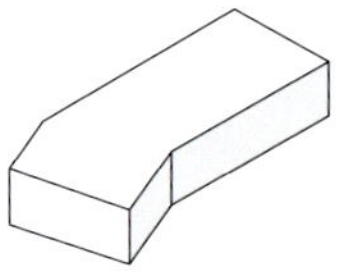

Octagon corner brick exterior

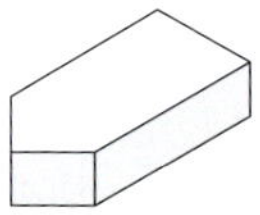

Octagon corner brick interior

Radial bricks

Radial bricks, with their gently curved surfaces, are used to create circular columns or serpentine walls. Radials form brickwork with smooth, sweeping curves, uninterrupted by sharp angles or jagged edges. Internal and external radial brick shapes, along with their companion starters, can be used to create details ranging from curtain-like walls to towering columns. These shapes are designed to form smooth, sweeping curves in brickwork without producing shadows or requiring field cuts.

Internal radial bricks feature a flat front and a curved internal edge—to the observer on the outside of the brickwork, a course of internal radial bricks would simply appear as a traditional wall or edge. On the other side, however, the curved areas would connect to form a smooth, circular angle. These bricks are used inside a curved wall for an even edge that doesn't require excessive mortar to connect.

External radial bricks provide a means to create curved edges ranging from gentle slopes in vertical walls to tighter, consciously-circular builds like institutional garden borders or decorative wishing wells. In order to determine the arc length necessary on an individual brick for circular projects, the overall width and length of the desired finished product can be multiplied and divided by 144.

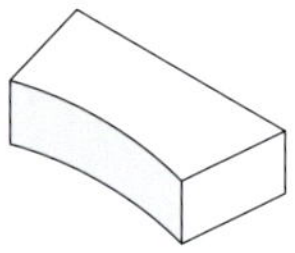

Internal radial brick

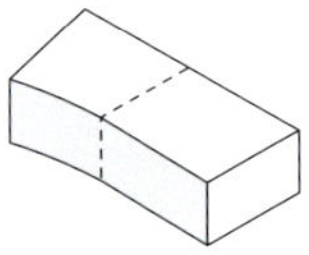

Internal radial brick - standard

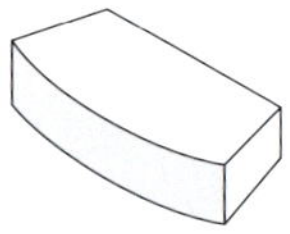

External radial brick

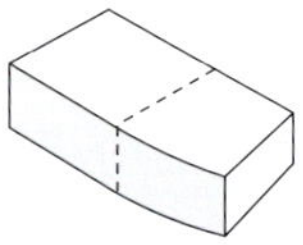

External radial brick - standard

Lipped bricks

Corner, lipped and radial brick shapes provide the masonry units necessary to achieve sound brickwork detailing while offering greater design flexibility. Lipped bricks are crafted with a hollow frog-like structure on their underside, serving a similar purpose to the coring details on hollow bricks. This space allows them to sit on a brickwork course evenly, with plenty of space for the mortar to flow in, settle, and dry. The end result is a smooth, even top edge to walls, borders, and decorative components of larger brickwork projects. The overhanging edge on lipped bricks, whether vertical or horizontally oriented, is usually standardized at ¾ in. for easy planning and incorporation.

When used with a fixed shelf angle, lipped bricks provide inconspicuous, horizontal expansion joints for brick masonry structures. They can also be used with loose lintels above window and door openings.

Flat arch bricks

Resembling a squared wedge shape (FA1-A) or a parallelogram (FA1-B) in style, flat arch bricks are used primarily to build brick masonry arches, either as standalone decorative components in a non-brick wall or incorporated as part of a larger brick border or building wall.

In some cases, these bricks may also be used as part of a brick sidewalk or pathway, used to introduce curvature, or used vertically at the ground level as decorative edging.

Channel brick

Channel bricks are molded to the shape of a gutter or a channel and are often glazed. These bricks are used as drains.

Custom shapes and sculptures

Some projects require a custom designed brick shape to achieve a specific detail or unusual effect. Some manufacturers work with the architect to create intricate design demands. This usually entails a high degree of skill and expertise to make and may take longer to produce than a standardized shape.

Sometimes, an artist creates a piece of brick sculpture. Depending on the manufacturer, sculpture can be designed to meet the requirements of the owner, the building or the environment, and can be executed in any size for interior and exterior walls or panels.

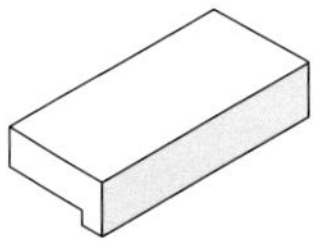

Lipped stretcher

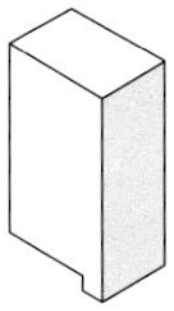

Lipped soldier

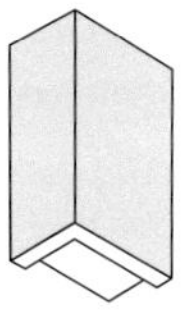

Lipped soldier corner - left or right option

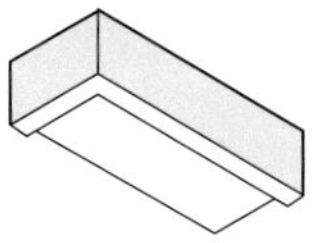

Lipped stretcher corner - left or right option

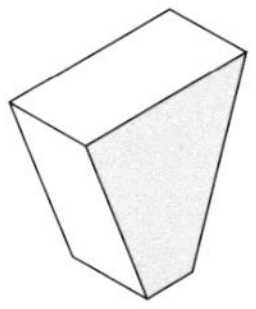

Flat arch keystone (FA1-A)

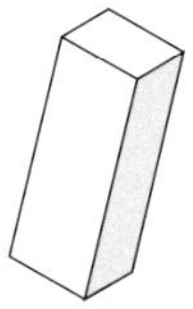

Flat arch voussoir (FA1-B)

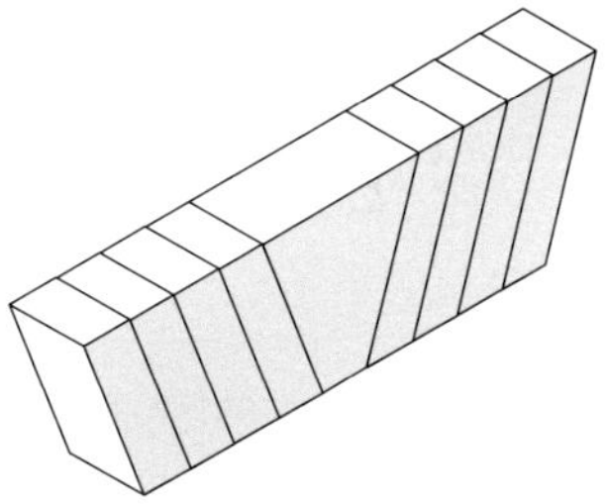

Flat arch

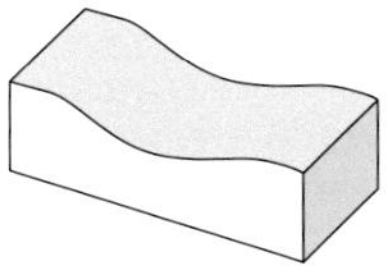

Channel brick

Methods of manufacture

There are several different types of brick and these are categorized in various ways depending on how they are manufactured.

Common clay bricks Dry-press bricks are formed by compressing the brick mixture into molds and then firing them in a kiln. Soft-mud bricks are made from a thin mix, whereas dry-press bricks are made from a thicker mix that gives crisper definition. Additional strength is achieved by using greater force when pressing the brick and by firing it for longer.

Common clay bricks can be categorized further as:

- First class—table-molded with smooth and straight surfaces
- Second class—ground-molded with rough and irregular surfaces; often used with plaster coating.
- Third class—ground-molded and used for temporary structures
- Fourth class—over-burnt with irregular shape used as aggregate for concrete in foundations, floors, etc

Sand lime bricks are manufactured by mixing sand, fly ash and lime. The brick is formed by molding the mix under pressure which achieves a smoother finish and a more uniform shape than clay bricks.

Engineering bricks are both dense and strong and manufactured at very high temperatures. This also means they have excellent load-bearing capacity, damp-proof and chemical-resistant properties. These advantages mean they are generally used for civil engineering applications such as groundworks, sewers, retaining walls, and so on.

Concrete bricks are made from solid concrete and can be used above and below damp-proof course level. They can also be manufactured in different colors if pigment is added during production.

Fly ash clay bricks This type of brick is manufactured with fly ash, cement, sand and water and fired at about 1,000 °C. Due to the high concentration of calcium oxide in fly ash, it is described as 'self-cementing'.

Historical feature

A wide variety of brick shapes, clay tiles and decorative pieces were available from brick and tile manufacturers until the late 1950s and 1960s. However, the cost of producing small runs of special or architecture bricks, as well as the impact of mass produced extruded standardized bricks resulted in falling demand for these pressed shapes. All of this further undermined the profitability of brick manufacturers making non-standard brick shapes.

In Australia, Central Brick & Tile Company Ltd at St Peters in Sydney manufactured huge quantities of brick, tiles and clay pieces. Unfortunately, in 1951 the factory was destroyed in a catastrophic fire. The examples right are taken from a promotional catalogue published in the late 1930s by the aforementioned firm. US manufacturers during this period would have produced an equally impressive number of brick specials and brick types.

Sources

The author wishes to acknowledge the use of information available from the following organizations.

ACME Brick - https://brick.com/special-brick-shapes

BigRentz – https://www.bigrentz.com/blog/brick-dimensions

Brick Architecture – https://brickarchitecture.com/about-brick/why-brick/brickwork-bonds

Civil Engineering – https://civiltoday.com/civil-engineering-materials/brick/271-brick-size-standard-dimensions

Designing Buildings: The Construction Wiki – https://www.designingbuildings.co.uk/wiki/Brick_sizes

General Shale – https://generalshale.com/custom-brick-shapes/

Glen-Gery – https://glengery.com/products/brick

Home Stratosphere – https://www.homestratosphere.com/types-of-bricks/

NCMA – https://ncma.org/resource/typical-sizes-and-shapes-of-concrete-masonry-units/

Old Carolina® Brick Company – https://www.handmadebrick.com/dir.cfm/BRICK_OPTIONS/SHAPES/

Triangle Brick – https://www.trianglebrick.com/resources/shapes-guide

US Brick – https://usbrick.com/shapes-and-sizes/

Wikipedia

Endnote

1 TMS 402/602-16 Building Code Requirements and Specification for Masonry Structures contains two standards and their commentaries: *Building Code Requirements for Masonry Structures (TMS 402-16)* and *Specification for Masonry Structures (TMS 602-16)*. The Standards are written as legal documents so that they may be adopted by reference in building codes. See https://masonrysociety.org/product/tms-402-602-2016/